Studies in Systems, Decision and Control

Volume 130

Series editor

Janusz Kacprzyk, Polish Academy of Sciences, Warsaw, Poland
e-mail: kacprzyk@ibspan.waw.pl

The series "Studies in Systems, Decision and Control" (SSDC) covers both new developments and advances, as well as the state of the art, in the various areas of broadly perceived systems, decision making and control- quickly, up to date and with a high quality. The intent is to cover the theory, applications, and perspectives on the state of the art and future developments relevant to systems, decision making, control, complex processes and related areas, as embedded in the fields of engineering, computer science, physics, economics, social and life sciences, as well as the paradigms and methodologies behind them. The series contains monographs, textbooks, lecture notes and edited volumes in systems, decision making and control spanning the areas of Cyber-Physical Systems, Autonomous Systems, Sensor Networks, Control Systems, Energy Systems, Automotive Systems, Biological Systems, Vehicular Networking and Connected Vehicles, Aerospace Systems, Automation, Manufacturing, Smart Grids, Nonlinear Systems, Power Systems, Robotics, Social Systems, Economic Systems and other. Of particular value to both the contributors and the readership are the short publication timeframe and the world-wide distribution and exposure which enable both a wide and rapid dissemination of research output.

More information about this series at http://www.springer.com/series/13304

George A. Anastassiou · Ioannis K. Argyros

Functional Numerical Methods: Applications to Abstract Fractional Calculus

 Springer

George A. Anastassiou
Department of Mathematical Sciences
University of Memphis
Memphis, TN
USA

Ioannis K. Argyros
Department of Mathematical Sciences
Cameron University
Lawton, OK
USA

ISSN 2198-4182   ISSN 2198-4190   (electronic)
Studies in Systems, Decision and Control
ISBN 978-3-319-88794-4   ISBN 978-3-319-69526-6   (eBook)
https://doi.org/10.1007/978-3-319-69526-6

Printed on acid-free paper

This Springer imprint is published by Springer Nature
The registered company is Springer International Publishing AG
The registered company address is: Gewerbestrasse 11, 6330 Cham, Switzerland

The first author dedicates this book to his wife Koula and daughters Angela and Peggy.

The second author dedicates this book to his wife Diana and children Christopher, Gus and Michael.

Preface

This is the abstract sequel booklet monograph to the recently published monographs, by the same authors, titled: "Intelligent Numerical Methods: Applications to Fractional Calculus," Studies in Computational Intelligence 624, and "Intelligent Numerical Methods II: Applications to Multivariate Fractional Calculus," Studies in Computational Intelligence 649, both in Springer Heidelberg New York, 2016. It is regarding applications of Newton-like and other similar methods for solving abstract functional equations, which involve abstract Caputo and Canavati type fractional derivatives. The functions we are dealing with are Banach space valued of a real domain. These are studied for the first time in the literature, and chapters are self-contained and can be read independently. In each chapter, the first sections are prerequisites for the final section of abstract fractional calculus applications. This short monograph is suitable to be used in related graduate classes and research projects. We exhibit the maximum of these numerical methods at the abstract fractional level.

The motivation to write this monograph came by the following: Various issues related to the modeling and analysis of fractional order systems have gained an increased popularity, as witnessed by many books and volumes in Springer's program:

http://www.springer.com/gp/search?query=fractional&submit=Prze%C5%9Blij

and the purpose of our book is to provide a deeper formal analysis on some issues that are relevant to many areas for instance: decision making, complex processes, systems modeling and control, and related areas. The above are deeply embedded in the fields of engineering, computer science, physics, economics, social and life sciences.

The list of covered topics here follows:

explicit–implicit methods with applications to Banach space valued functions in abstract fractional calculus,
convergence of iterative methods in abstract fractional calculus,
equations for Banach space valued functions in fractional vector calculi,

iterative methods in abstract fractional calculus,
semi-local convergence in right abstract fractional calculus,
algorithmic convergence in abstract g-fractional calculus,
iterative procedures for solving equations in abstract fractional calculus,
approximate solutions of equations in abstract g-fractional calculus,
generating sequences for solving equations in abstract g-fractional calculus,
and numerical optimization with fractional invexity.

An extensive list of references is given per chapter.

This book's results are expected to find applications in many areas of applied mathematics, stochastics, computer science, and engineering. As such, this short monograph is suitable for researchers, graduate students, and seminars of the above subjects, also to be in all science and engineering libraries.

The preparation of this book took place during the academic year 2016–2017 in Memphis, Tennessee, and Lawton, Oklahoma, USA.

We would like to thank Prof. Alina Lupas of University of Oradea, Romania, for checking and reading the manuscript.

Memphis, USA   George A. Anastassiou
Lawton, USA   Ioannis K. Argyros
June 2017

Contents

Chapter 1
Explicit-Implicit Methods with Applications to Banach Space Valued Functions in Abstract Fractional Calculus

Explicit iterative methods have been used extensively to generate a sequence approximating a solution of an equation on a Banach space setting. However, little attention has been given to the study of implicit iterative methods. We present a semi-local convergence analysis for a some general implicit and explicit iterative methods. Some applications are suggested including Banach space valued functions of fractional calculus, where all integrals are of Bochner-type. It follows [5].

1.1 Introduction

Sections 1.1–1.3 are prerequisites for Sect. 1.4.

Let B_1, B_2 stand for Banach spaces and let Ω stand for an open subset of B_1. Let also $U(z, \rho) := \{u \in B_1 : \|u - z\| < \rho\}$ and let $\overline{U}(z, \rho)$ stand for the closure of $U(z, \rho)$.

Numerous problems in Computational Sciences, Engineering, Mathematical Chemistry, Mathematical Physics, Mathematical Economics and other disciplines can be brought in a form like

$$F(x) = 0 \qquad (1.1.1)$$

using Mathematical Modeling [1–16], where $F : \Omega \to B_2$ is a continuous operator. The solution x^* of Eq. (1.1.1) is sought in closed form, but this is attainable only in special cases. That explains why most solution methods for such equations are usually iterative. There is a plethora of iterative methods for solving Eq. (1.1.1). We can divide these methods in two categories.

Explicit Methods [6, 7, 11, 14, 15]: Newton's method

$$x_{n+1} = x_n - F'(x_n)^{-1} F(x_n). \qquad (1.1.2)$$

© Springer International Publishing AG 2018
G. A. Anastassiou and I. K. Argyros, *Functional Numerical Methods: Applications to Abstract Fractional Calculus*, Studies in Systems, Decision and Control 130, https://doi.org/10.1007/978-3-319-69526-6_1

Secant method:

$$x_{n+1} = x_n - [x_{n-1}, x_n; F]^{-1} F(x_n), \tag{1.1.3}$$

where $[\cdot, \cdot; F]$ denotes a divided difference of order one on $\Omega \times \Omega$ [7, 14, 15].

Newton-like method:

$$x_{n+1} = x_n - E_n^{-1} F(x_n), \tag{1.1.4}$$

where $E_n = E(F)(x_n)$ and $E : \Omega \to \mathcal{L}(B_1, B_2)$ the space of bounded linear operators from B_1 into B_2. Other explicit methods can be found in [7, 11, 14, 15] and the references there in.

Implicit Methods [6, 9, 11, 15]:

$$F(x_n) + A_n(x_{n+1} - x_n) = 0 \tag{1.1.5}$$

$$x_{n+1} = x_n - A_n^{-1} F(x_n), \tag{1.1.6}$$

where $A_n = A(x_{n+1}, x_n) = A(F)(x_{n+1}, x_n)$ and $A : \Omega \times \Omega \to \mathcal{L}(B_1, B_2)$.

There is a plethora on local as well as semi-local convergence results for explicit methods [1–8], [10–15]. However, the research on the convergence of implicit methods has received little attention. Authors, usually consider the fixed point problem

$$P_z(x) = x, \tag{1.1.7}$$

where

$$P_z(x) = x + F(z) + A(x, z)(x - z) \tag{1.1.8}$$

or

$$P_z(x) = z - A(x, z)^{-1} F(z) \tag{1.1.9}$$

for methods (1.1.5) and (1.1.6), respectivelly, where $z \in \Omega$ is given. If P is a contraction operator mapping a closed set into itself, then according to the contraction mapping principle [11, 14, 15], P_z has a fixed point x_z^* which can be found using the method of succesive substitutions or Picard's method [15] defined for each fixed n by

$$y_{k+1,n} = P_{x_n}(y_{k,n}), \quad y_{0,n} = x_n, \ x_{n+1} = \lim_{k \to +\infty} y_{k,n}. \tag{1.1.10}$$

Let us also consider the analogous explicit methods

$$F(x_n) + A(x_n, x_n)(x_{n+1} - x_n) = 0 \tag{1.1.11}$$

$$x_{n+1} = x_n - A(x_n, x_n)^{-1} F(x_n) \tag{1.1.12}$$

$$F(x_n) + A(x_n, x_{n-1})(x_{n+1} - x_n) = 0 \tag{1.1.13}$$

and

$$x_{n+1} = x_n - A(x_n, x_{n-1})^{-1} F(x_n). \tag{1.1.14}$$

In this chapter in Sect. 1.2, we study the semi-local convergence of method (1.1.5) and method (1.1.6). Section 1.3 contains the semi-local convergence of method (1.1.11), method (1.1.12), method (1.1.13) and method (1.1.14). Some applications to Abstract Fractional Calculus are suggested in Sect. 1.4 on a certain Banach space valued functions, where all the integrals are of Bochner-type [8].

1.2 Semi-local Convergence for Implicit Methods

The semi-local convergence analysis of method (1.1.6) that follows is based on the conditions (H):

(h_1) $F : \Omega \subset B_1 \to B_2$ is continuous and $A(F)(x, y) \in \mathcal{L}(B_1, B_2)$ for each $(x, y) \in \Omega \times \Omega$.

(h_2) There exist $l > 0$ and $\Omega_0 \subset B_1$ such that $A(F)(x, y)^{-1} \in \mathcal{L}(B_2, B_1)$ for each $(x, y) \in \Omega_0 \times \Omega_0$ and

$$\left\| A(F)(x, y)^{-1} \right\| \le l^{-1}.$$

Set $\Omega_1 = \Omega \cap \Omega_0$.

(h_3) There exist real numbers $\alpha_1, \alpha_2, \alpha_3$ satisfying

$$0 \le \alpha_2 \le \alpha_1 \text{ and } 0 \le \alpha_3 < 1$$

such that for each $x, y \in \Omega_1$

$$\| F(x) - F(y) - A(F)(x, y)(x - y) \| \le$$

$$l \left(\frac{\alpha_1}{2} \|x - y\| + \alpha_2 \|y - x_0\| + \alpha_3 \right) \|x - y\|.$$

(h_4) For each $x \in \Omega_0$ there exists $y \in \Omega_0$ such that

$$y = x - A(y, x)^{-1} F(x).$$

(h_5) For $x_0 \in \Omega_0$ and $x_1 \in \Omega_0$ satisfying (h_4) there exists $\eta \ge 0$ such that

$$\left\| A(F)(x_1, x_0)^{-1} F(x_0) \right\| \le \eta.$$

(h_6) $h := \alpha_1 \eta \le \frac{1}{2}(1 - \alpha_3)^2$.
and
(h_7) $\overline{U}(x_0, t^*) \subset \Omega_0$, where

$$t^* = \begin{cases} \frac{1-\alpha_3-\sqrt{(1-\alpha_3)^2-2h}}{\alpha_1}, & \alpha_1 \neq 0 \\ \frac{1}{1-\alpha_3}\eta, & \alpha_1 = 0. \end{cases}$$

Then, we can show the following semi-local convergence result for method (1.1.6) under the preceding notation and conditions (H).

Theorem 1.1 *Suppose that the conditions (H) are satisfied. Then, sequence $\{x_n\}$ generated by method (1.1.6) starting at $x_0 \in \Omega$ is well defined in $U(x_0, t^*)$, remains in $U(x_0, t^*)$ for each $n = 0, 1, 2, ...$ and converges to a solution $x^* \in \overline{U}(x_0, t^*)$ of equation $F(x) = 0$. Moreover, provided that (h_3) holds with $A(F)(z,y)$ replacing $A(F)(x,y)$ for each $z \in \Omega_1$, if $\alpha_1 \neq 0$, the equation $F(x) = 0$ has a unique solution x^* in $\overline{U}$, where*

$$\widetilde{U} = \begin{cases} \overline{U}(x_0, t^*) \cap \Omega_0, & \text{if } h = \frac{1}{2}(1-\alpha_3)^2 \\ U(x_0, t^{**}) \cap \Omega_0, & \text{if } h < \frac{1}{2}(1-\alpha_3)^2 \end{cases}$$

and, if $\alpha_1 = 0$, the solution x^* is unique in $\overline{U}\left(x_0, \frac{\eta}{1-\alpha_3}\right)$, where $t^{**} = \frac{1-\alpha_3+\sqrt{(1-\alpha_3)^2-2h}}{\alpha_1}$.

Proof Case $\alpha_1 \neq 0$. Let us define scalar function g on $\mathbb{R}$ by $g(t) = \frac{\alpha_1}{2}t^2 - (1-\alpha_3)t + \eta$ and majorizing sequence $\{t_n\}$ by

$$t_0 = 0, \quad t_k = t_{k-1} + g(t_{k-1}) \quad \text{for each } k = 1, 2, ... \tag{1.2.1}$$

It follows from (h_6) that function g has two positive roots t^* and t^{**}, $t^* \leq t^{**}$, and $t_k \leq t_{k+1}$. That is, sequence $\{t_k\}$ converges to t^*.

(a) Using mathematical induction on k, we shall show that

$$\|x_{k+1} - x_k\| \leq t_{k+1} - t_k. \tag{1.2.2}$$

Estimate (1.2.2) holds for $k = 0$ by (h_5) and (1.2.1), since $\|x_1 - x_0\| \leq \eta = t_1 - t_0$. Suppose that for $1 \leq m \leq k$

$$\|x_m - x_{m-1}\| \leq t_m - t_{m-1}. \tag{1.2.3}$$

Them, we get $\|x_k - x_0\| \leq t_k - t_0 = t_k \leq t^*$ and $A(x_k, x_{k-1})$ is invertible by (h_2). We can write by method (1.1.6)

$$x_{k+1} - x_k = -A_k^{-1}(F(x_k) - F(x_{k-1}) - A_{k-1}(x_k - x_{k-1})). \tag{1.2.4}$$

In view of the induction hypothesis (1.2.3), (h_2), (h_3), (h_4), (1.2.1) and (1.2.4), we get in turn that

$$\|x_{k+1} - x_k\| = \left\|A_k^{-1}F(x_k)\right\| = \left\|A_k^{-1}(F(x_k) - F(x_{k-1}) - A_{k-1}(x_k - x_{k-1}))\right\|$$

$$\leq \left\| A_k^{-1} \right\| \left\| F(x_k) - F(x_{k-1}) - A_{k-1}(x_k - x_{k-1}) \right\| \leq$$

$$l^{-1} l \left(\frac{\alpha_1}{2} \left\| x_k - x_{k-1} \right\| + \alpha_2 \left\| x_{k-1} - x_0 \right\| + \alpha_3 \right) \left\| x_k - x_{k-1} \right\| \leq \tag{1.2.5}$$

$$\frac{\alpha_1}{2} (t_k - t_{k-1})^2 + \alpha_2 (t_k - t_{k-1}) t_{k-1} + \alpha_3 (t_k - t_{k-1}) =$$

$$\frac{\alpha_1}{2} (t_k - t_{k-1})^2 + \alpha_2 (t_k - t_{k-1}) t_{k-1} + \alpha_3 (t_k - t_{k-1}) - (t_k - t_{k-1}) + g(t_{k-1}) =$$

$$g(t_k) - (\alpha_1 - \alpha_2) (t_k - t_{k-1}) t_{k-1} \leq$$

$$g(t_k) = t_{k+1} - t_k, \tag{1.2.6}$$

which completes the induction for estimate (1.2.2).

That is, we have for any k

$$\left\| x_{k+1} - x_k \right\| \leq t_{k+1} - t_k \tag{1.2.7}$$

and

$$\left\| x_k - x_0 \right\| \leq t_k \leq t^*. \tag{1.2.8}$$

It follows by (1.2.7) and (1.2.8) that $\{x_k\}$ is a complete sequence in a Banach space B_1 and as such it converges to some $x^* \in \overline{U}(x_0, t^*)$ (since $\overline{U}(x_0, t^*)$ is a closed set). By letting $k \to +\infty$, using (h_1) and (h_2), we get $l^{-1} \lim\limits_{k \to +\infty} \|F(x_k)\| = 0$, so $F(x^*) = 0$.

Let $x^{**} \in \widetilde{U}$ be such that $F(x^{**}) = 0$. We shall show by induction that

$$\left\| x^{**} - x_k \right\| \leq t^* - t_k \quad \text{for each } k = 0, 1, 2, \ldots . \tag{1.2.9}$$

Estimate (1.2.9) holds for $k = 0$ by the definition of x^{**} and $\widetilde{U}$. Suppose that $\|x^{**} - x_k\| \leq t^* - t_k$. Then, as in (1.2.5), we obtain in turn that

$$\left\| x^{**} - x_{k+1} \right\| = \left\| x^{**} - x_k + A_k^{-1} F(x_k) - A_k^{-1} F(x^{**}) \right\| =$$

$$\left\| A_k^{-1} \left(A_k (x^{**} - x_k) + F(x_k) - F(x^{**}) \right) \right\| \leq$$

$$\left\| A_k^{-1} \right\| \left\| F(x^{**}) - F(x_k) - A_k (x^{**} - x_k) \right\| \leq$$

$$\left(\frac{\alpha_1}{2} \left\| x^{**} - x_k \right\| + \alpha_2 \left\| x_k - x_0 \right\| + \alpha_3 \right) \left\| x^{**} - x_k \right\| \leq$$

$$\left(\frac{\alpha_1}{2} (t^* - t_k) + \alpha_2 t_k + \alpha_3 \right) (t^* - t_k) =$$

$$\frac{\alpha_1}{2} \left(t^*\right)^2 + \frac{\alpha_1}{2} \left(t_k\right)^2 - \alpha_1 t_k t^* + \alpha_2 \left(t^* - t_k\right) t_k + \alpha_3 \left(t^* - t_k\right) =$$

$$-\eta + (1 - \alpha_3) t^* + \frac{\alpha_1}{2} t_k^2 - \alpha_1 t_k t^* + \alpha_2 t_k t^* - \alpha_2 t_k^2 + \alpha_3 t^* - \alpha_3 t_k$$

$$= t^* - t_{k+1}, \tag{1.2.10}$$

which completes the induction for estimate (1.2.9). Hence, $\lim_{k \to +\infty} x_k = x^{**}$. But we showed that $\lim_{k \to +\infty} x_k = x^*$, so $x^{**} = x^*$.

Case $\alpha_1 = 0$. Then, we have by (h_3) that $\alpha_2 = 0$ and estimate (1.2.5) gives

$$\|x_{k+1} - x_k\| \le \alpha_3 \|x_k - x_{k-1}\| \le \cdots \le \alpha_3^k \|x_1 - x_0\| \le \alpha_3^k \eta \tag{1.2.11}$$

and

$$\|x_{k+1} - x_0\| \le \|x_{k+1} - x_k\| + \|x_k - x_{k-1}\| + \cdots + \|x_1 - x_0\|$$

$$\le \frac{1 - \alpha_3^{k+1}}{1 - \alpha_3} \eta < \frac{\eta}{1 - \alpha_3}. \tag{1.2.12}$$

Then, as in the previous case it follows from (1.2.11) and (1.2.12) that

$$\|x_{k+i} - x_k\| \le \frac{1 - \alpha_3^i}{1 - \alpha_3} \alpha_3^k \eta, \tag{1.2.13}$$

so sequence $\{x_k\}$ is complete and x^* solves equation $F(x) = 0$. Finally, the uniqueness part follows from (1.2.10) for $\alpha_1 = \alpha_2 = 0$, since

$$\left\|x^{**} - x_{k+1}\right\| \le \alpha_3 \left\|x^{**} - x_k\right\| \le \alpha_3^{k+1} \left\|x^{**} - x_0\right\| \le \alpha_3^{k+1} \frac{\eta}{1 - \alpha_3}, \tag{1.2.14}$$

which shows again that $\lim_{k \to +\infty} x_k = x^{**}$.   ■

Remark 1.2 (1) Condition (h_2) can be incorporated in (h_3) as follows

(h_3') There exist real numbers $\overline{\alpha}_1, \overline{\alpha}_2, \overline{\alpha}_3$ satisfying $0 \le \overline{\alpha}_2 \le \overline{\alpha}_1$ and $0 \le \overline{\alpha}_3 < 1$ such that for each $x, y \in \Omega$

$$\left\| A(x, y)^{-1} \left[F(x) - F(y) - A(x, y)(x - y) \right] \right\| \le$$

$$\left((\overline{\alpha}_1/2) \|x - y\| + \overline{\alpha}_2 \|y - x_0\| + \overline{\alpha}_3 \right) \|x - y\|.$$

Then, (h_3') can replace (h_2) and (h_3) in Theorem 1.1 for $\alpha_1 = \overline{\alpha}_1, \overline{\alpha}_2 = \alpha_2$, $\overline{\alpha}_3 = \alpha_3$ and $\Omega_0 = \Omega$. Moreover, notice that $\overline{\alpha}_1 \le \alpha_1, \overline{\alpha}_2 \le \alpha_1$ and $\overline{\alpha}_3 \le \alpha_3$, which play a role in the sufficient convergence criterion (h_6), error bounds and the precision of t^* and t^{**}. Condition (h_3) is of Mysowksii-type [11].

(2) Suppose that there exist $l_0 > 0$, $\alpha_4 > 0$ and $L \in \mathcal{L}(B_1, B_2)$ with $L^{-1} \in \mathcal{L}(B_2, B_1)$ such that $\left\| L^{-1} \right\| \leq l_0^{-1}$

$$\| A(F)(x, y) - L \| \leq \alpha_4 \text{ for each } x, y \in \Omega$$

and

$$\alpha_5 := l_0^{-1} \alpha_4 < 1.$$

Then, it follows from the Banach lemma on invertible operators [7, 9, 11, 14, 15] and

$$\left\| L^{-1} \right\| \| A(F)(x, y) - L \| \leq l_0^{-1} \alpha_4 = \alpha_5 < 1$$

that $A(F)(x, y)^{-1} \in \mathcal{L}(B_2, B_1)$. Set $l^{-1} = \frac{l_0^{-1}}{1 - \alpha_5}$. Then, under these replacements, condition (h_2) is implied, so it can be dropped from the conditions (H).

(3) Clearly method (1.1.5) converges under the conditions (H), since (1.1.6) implies (1.1.5).

(4) Let $R > 0$ and define $R_0 = \sup\{t \in [0, R) : U(x_0, R_0) \subseteq D\}$. Set $\Omega_0 = \overline{U}(x_0, R_0)$. Condition (h_3) can be extended, if the additional term $a_2 \|x - x_0\|$ is inserted inside the paranthesis at the right hand side for some $a_2 \geq 0$. Then, the conclusions of Theorem 1.1 hold in this more general setting, provided that $a_3 = a_2 R_0 + \alpha_3$ replaces α_3 in conditions (h_6) and (h_7).

(5) Concerning the solvability of Eq. (1.1.6) (or (1.1.5)), we wanted to leave condition (h_4) as uncluttered as possible in conditions (H). We did this because in practice these equations may be solvable in a way other than using the contraction mapping principle already mentioned earlier.

Next, we show the solvability of method (1.1.5) using a stronger version of the contraction mapping principle and based on the conditions (C) :

$(c_1) = (h_1)$.

(c_2) There exist $\gamma_0 \in [0, 1)$, $\gamma_1 \in [0, +\infty)$, $\gamma_2 \in [0, 1)$, $x_0 \in \Omega$ such that for each $x, y, z \in \Omega$

$$\| I + A(x, z) - A(y, z) \| \leq \gamma_0,$$

$$\| A(x, z) - A(y, z) \| \leq \gamma_1 \|x - y\|$$

$$\| F(z) + A(x_0, z)(x_0 - z) \| \leq \begin{cases} \gamma_2 \|x_0 - z\| & \text{for } x_0 \neq z \\ \| F(x_0) \| & \text{for } x_0 = z \end{cases}$$

(c_3)

$$\gamma_0 + \gamma_1 \|x_0\| + \gamma_2 \leq 1 \text{ for } \gamma_2 \neq 0,$$

$$\gamma_0 + \gamma_1 \|x_0\| < 1 \text{ for } \gamma_2 = 0,$$

$$\| F(x_0) \| \leq \frac{(1 - (\gamma_0 + \gamma_1 \|x_0\|))^2}{\gamma_1} \text{ for } \gamma_1 \neq 0,$$

$$\gamma_0 < 1 \text{ for } \gamma_1 = 0$$

and

(c_4) $\overline{U}(x_0, r) \subseteq \Omega$, where

$$\frac{\|F(x_0)\|}{1 - (\gamma_0 + \gamma_1 \|x_0\|)} \leq r < \frac{1 - (\gamma_0 + \gamma_1 \|x_0\|)}{\gamma_1} \text{ for } \gamma_1 \neq 0,$$

$$\frac{\|F(x_0)\|}{1 - \gamma_0} \leq r \text{ for } \gamma_1 = 0,$$

$$r < \frac{1 - (\gamma_0 + \gamma_1 \|x_0\|)}{\gamma_1} \text{ for } z = x_0, \gamma_1 \neq 0.$$

Theorem 1.3 *Suppose that the conditions (C) are satisfied. Then, for each $n = 0, 1, 2, ...$ Eq. (1.1.5) is unique solvable. Moreover, if $A_n^{-1} \in \mathcal{L}(B_2, B_1)$, then Eq. (1.1.6) is also uniquely solvable for each $n = 0, 1, 2, ...$*

Proof We base the proof on the contraction mapping principle. Let $x, y \in U(x_0, r)$. Then, using (1.1.8) we have in turn by (c_2) that

$$\|P_z(x) - P_z(y)\| = \|(I + A(x, z) - A(y, z))(x - y) - (A(x, z) - A(y, z))z\|$$

$$\leq \|I + A(x, z) - A(y, z)\| \|x - y\| + \|A(x, z) - A(y, z)\| \|z\|$$

$$\leq \gamma_0 \|x - y\| + \gamma_1 (\|z - x_0\| + \|x_0\|) \|x - y\|$$

$$\leq \varphi(\|x - x_0\|) \|x - y\|, \tag{1.2.15}$$

where

$$\varphi(t) = \begin{cases} \gamma_0 + \gamma_1 (t + \|x_0\|) & \text{for } z \neq x_0 \\ \gamma_0 + \gamma_1 \|x_0\| & \text{for } z = x_0. \end{cases} \tag{1.2.16}$$

Notice that $\varphi(t) \in [0, 1)$ for $t \in [0, r]$ by the choice of r in (c_4).

We also have that

$$\|P_z(x) - x_0\| \leq \|P_z(x) - P_z(x_0)\| + \|P_z(x_0) - x_0\|. \tag{1.2.17}$$

If $z = x_0$ in (1.2.17), then we get by (c_3), (c_4) and (1.2.15) that

$$\left\| P_{x_0}(x) - x_0 \right\| \leq \varphi(\|x - x_0\|) \|x - x_0\| + \|F(x_0)\|$$

$$\leq (\gamma_0 + \gamma_1 \|x_0\|) r + \|F(x_0)\| \leq r. \tag{1.2.18}$$

The existence of $x_1 \in U(x_0, r)$ solving (1.1.5) for $n = 0$ is now established by the contraction mapping principle, (1.2.15) and (1.2.18).

Moreover, if $z \neq x_0$, the last condition in (c_3), (c_3), (c_4) and (1.2.17) give instead of (1.2.18) that

$$\|P_z(x) - x_0\| \leq \varphi(\|x - x_0\|)\|x - x_0\| + \gamma_2\|x - x_0\|$$

$$\leq (\gamma_0 + \gamma_1\|x_0\| + \gamma_2)r \leq r. \tag{1.2.19}$$

Then, again by (1.2.15), (1.2.19) and the contraction mapping principle, we guarantee the unique solvability of Eq. (1.1.5) and the existence of a unique sequence $\{x_n\}$ for each $n = 0, 1, 2, \ldots$ Finally, equation (1.1.6) is also uniquely solvable by the preceding proof and the condition $A_n^{-1} \in \mathcal{L}(B_2, B_1)$. ∎

Remark 1.4 (a) The gamma conditions can be weakened, if γ_i are replaced by functions $\gamma_i(t)$, $i = 0, 1, 2, 3$. Then, γ_i will appear as $\gamma_i(\|x - x_0\|)$ and $\gamma_i(r)$ in the conditions (C).

(b) Sect. 1.2 has an interest independent of Sect. 1.4. However, the results especially of Theorem 1.1 can apply in Abstract Fractional Calculus as suggested in Sect. 1.4. As an example crucial condition (h_3) is satisfied in (1.4.8), if we choose $\alpha_2 = \alpha_3 = 0$ and $l\alpha_1 = \frac{c}{2}$, where c is defined in (1.4.8). Similar choices can be given for the rest of the special cases of (h_3) appearing in Sect. 1.4.

1.3 Semi-local Convergence for Explicit Methods

Theorem 1.1 is general enough so it can be used to study the semi-local convergence of method (1.1.11), method (1.1.12), method (1.1.13) and method (1.1.14). In particular, for the study of method (1.1.12) (and consequently method (1.1.11)), we use the conditions (H') :

(h'_1) $F : \Omega \subset B_1 \rightarrow B_2$ is continuous and $A(F)(x, x) \in \mathcal{L}(B_1, B_2)$ for each $x \in \Omega$.

(h'_2) There exist $l > 0$ and $\Omega_0 \subset B_1$ such that $A(F)(x, x)^{-1} \in \mathcal{L}(B_2, B_1)$ and

$$\left\|A(F)(x, x)^{-1}\right\| \leq l^{-1}.$$

Set $\Omega_1 = \Omega \cap \Omega_0$.

(h'_3) There exist real numbers $\gamma_1, \alpha_2, \gamma_3$ satisfying

$$0 \leq \alpha_2 \leq \gamma_1 \text{ and } 0 \leq \gamma_3$$

such that for each $x, y \in \Omega_1$

$$\|F(x) - F(y) - A(F)(y, y)(x - y)\| \leq$$

$$l\left(\frac{\gamma_1}{2}\|x - y\| + \alpha_2\|y - x_0\| + \gamma_3\right)\|x - y\|.$$

(h'_4) For each $x, y \in \Omega_1$ and some $\gamma_4 \geq 0$, $\gamma_5 \geq 0$

$$\|A(x, y) - A(y, y)\| \leq l\gamma_4$$

or

$$\|A(x, y) - A(y, y)\| \leq l\gamma_5 \|x - y\|.$$

Set $\alpha_1 = \gamma_1 + \gamma_5$ and $\alpha_3 = \gamma_3 + \gamma_4$, if the second inequation holds or $\alpha_1 = \gamma_1$ and $\alpha_3 = \gamma_3 + \gamma_4$, if the first inequation holds. Further, suppose $0 \leq \alpha_3 < 1$.

(h'_5) There exist $x_0 \in \Omega_0$ and $\eta \geq 0$ such that $A(F)(x_0, x_0)^{-1} \in \mathcal{L}(B_2, B_1)$ and

$$\left\|A(F)(x_0, x_0)^{-1} F(x_0)\right\| \leq \eta.$$

$(h'_6) = (h_6)$
and
$(h'_7) = (h_7)$.

Then, we can show the following semi-local convergence of method (1.1.12) using the conditions (H') and the preceding notation.

Proposition 1.5 *Suppose that the conditions (H') are satisfied. Then, sequence $\{x_n\}$ generated by method (1.1.12) starting at $x_0 \in \Omega$ is well defined in $U(x_0, t^*)$, remains in $U(x_0, t^*)$ for each $n = 0, 1, 2, \ldots$ and converges to a solution $x^* \in \overline{U}(x_0, t^*)$ of equation $F(x) = 0$. Moreover, if $\alpha_1 \neq 0$, the equation $F(x) = 0$ has a unique solution x^* in $\widetilde{U}$, where*

$$\widetilde{U} = \begin{cases} \overline{U}(x_0, t^*) \cap \Omega_0, & \text{if } h = \frac{1}{2}(1 - \alpha_3)^2 \\ U(x_0, t^{**}) \cap \Omega_0, & \text{if } h < \frac{1}{2}(1 - \alpha_3)^2 \end{cases}$$

and, if $\alpha_1 = 0$, the solution x^* is unique in $\overline{U}\left(x_0, \frac{\eta}{1-\alpha_3}\right)$, where t^* and t^{**} are given in Theorem 1.1.

Proof Use in the proof of Theorem 1.1 instead of estimate (1.2.5) the analogous estimate

$$\|F(x_k)\| = \|F(x_k) - F(x_{k-1}) - A(x_{k-1}, x_{k-1})(x_k - x_{k-1})\| =$$

$$\left\|\left[F(x_k) - F(x_{k-1}) - A(x_k, x_{k-1})(x_k - x_{k-1})\right] +\right.$$

$$\left.(A(x_k, x_{k-1}) - A(x_{k-1}, x_{k-1}))(x_k - x_{k-1})\right\|$$

$$\leq l\left(\frac{\gamma_1}{2}\|x_k - x_{k-1}\| + \alpha_2\|x_{k-1} - x_0\| + \gamma_3\right)\|x_k - x_{k-1}\| +$$

$$\|A(x_k, x_{k-1}) - A(x_{k-1}, x_{k-1})\|\|x_k - x_{k-1}\| \leq$$

$$l\left(\frac{\alpha_1}{2}(t_k - t_{k-1})^2 + \alpha_2 (t_k - t_{k-1}) t_{k-1} + \alpha_3 (t_k - t_{k-1})\right),$$

where we used again that $\|x_k - x_{k-1}\| \le t_k - t_{k-1}$, $\|x_{k-1} - x_0\| \le t_{k-1}$ and the condition (h'_4). ∎

Remark 1.6 Comments similar to Remark 1.2 (1)–(3) can follow but for method (1.1.11) and method (1.1.12) instead of method (1.1.5) and method (1.1.6), respectively.

Similarly, for method (1.1.13) and method (1.1.14), we use the conditions (H'') :
$(h''_1) = (h_1)$
$(h''_2) = (h_2)$
(h''_3) There exist real numbers $\alpha_1, \alpha_2, \gamma_3$ satisfying

$$0 \le \alpha_2 \le \alpha_1 \text{ and } 0 \le \gamma_3$$

such that for each $x, y \in \Omega_1$

$$\|F(x) - F(y) - A(F)(x, y)(x - y)\| \le$$

$$l\left(\frac{\alpha_1}{2}\|x - y\| + \alpha_2 \|y - x_0\| + \gamma_3\right) \|x - y\|.$$

(h''_4) For each $x, y, z \in \Omega_1$ and some $\gamma_3 \ge 0$

$$\|A(z, y) - A(y, x)\| \le l\delta_3.$$

Set $\alpha_3 = \gamma_3 + \delta_3$ and further suppose $0 \le \alpha_3 < 1$.
(h''_5) There exist $x_{-1} \in \Omega$, $x_0 \in \Omega$ and $\eta \ge 0$ such that $A(F)(x_0, x_{-1})^{-1} \in \mathcal{L}(B_2, B_1)$ and

$$\|A(F)(x_0, x_{-1})^{-1} F(x_0)\| \le \eta.$$

$(h''_6) = (h_6)$
and
$(h''_7) = (h_7)$.
Then, we can present the following semi-local convergence of method (1.1.14) using the conditions (H'') and the preceding notation.

Proposition 1.7 *Suppose that the conditions (H'') are satisfied. Then, sequence $\{x_n\}$ generated by method (1.1.14) starting at $x_0 \in \Omega$ is well defined in $U(x_0, t^*)$, remains in $U(x_0, t^*)$ for each $n = 0, 1, 2, \dots$ and converges to a solution $x^* \in \overline{U}(x_0, t^*)$ of equation $F(x) = 0$. Moreover, if $\alpha_1 \ne 0$, the equation $F(x) = 0$ has a unique solution x^* in $\tilde{U}$, where*

$$\tilde{U} = \begin{cases} \overline{U}(x_0, t^{**}) \cap \Omega_0, & \text{if } h = \frac{1}{2}(1 - \alpha_3)^2 \\ U(x_0, t^{**}) \cap \Omega_0, & \text{if } h < \frac{1}{2}(1 - \alpha_3)^2 \end{cases}$$

and, if $\alpha_1 = 0$, the solution x^* is unique in $\overline{U}\left(x_0, \frac{\eta}{1-\alpha_3}\right)$, where t^* and t^{**} are given in Theorem 1.1.

Proof As in Proposition 1.5, use in the proof of Theorem 1.1 instead of estimate (1.2.5) the analogous estimate

$$\|F(x_k)\| =$$

$$\|F(x_k) - F(x_{k-1}) - A(x_k, x_{k-1})(x_k - x_{k-1})$$

$$+ (A(x_k, x_{k-1}) - A(x_{k-1}, x_{k-2}))(x_k - x_{k-1})\| \le$$

$$\|F(x_k) - F(x_{k-1}) - A(x_k, x_{k-1})(x_k - x_{k-1})\| +$$

$$\|A(x_k, x_{k-1}) - A(x_{k-1}, x_{k-2})\| \, \|x_k - x_{k-1}\|$$

$$\le l\left(\frac{\alpha_1}{2}\|x_k - x_{k-1}\| + \alpha_2\|x_{k-1} - x_0\| + \gamma_3\right)\|x_k - x_{k-1}\| + l\delta_3\|x_k - x_{k-1}\|$$

$$\le l\left(\frac{\alpha_1}{2}(t_k - t_{k-1})^2 + \alpha_2(t_k - t_{k-1})t_{k-1} + \alpha_3(t_k - t_{k-1})\right),$$

where we used again that $\|x_k - x_{k-1}\| \le t_k - t_{k-1}$, $\|x_{k-1} - x_0\| \le t_{k-1}$ and the condition $\left(h_4''\right)$. ∎

Remark 1.8 Comments similar to Remark 1.2 (1)–(3) can follow but for method (1.1.13) and method (1.1.14) instead of method (1.1.5) and method (1.1.6), respectively.

1.4 Applications to X-valued Fractional Calculus

Here we deal with Banach space $(X, \|\cdot\|)$ valued functions f of real domain $[0, a]$, $a > 0$. All integrals here are of Bochner-type, see [8, 13]. The derivatives of f are defined similarly to numerical ones, see [16], pp. 83–86 and p. 93.

In this section we apply our Newton like numerical methods to X-valued fractional calculus.

We want to solve

$$f(x) = 0. \tag{1.4.1}$$

(I) Let $1 < \nu < 2$, i.e. $\lceil \nu \rceil = 2$ ($\lceil \cdot \rceil$ ceiling of number); $x, y \in [0, a]$, $a > 0$, and $f \in C^2([0, a], X)$.

We define the following left X-valued Caputo fractional derivatives (see [4]),

$$\left(D_{*y}^{\nu} f\right)(x) := \frac{1}{\Gamma(2-\nu)} \int_y^x (x-t)^{1-\nu} f''(t)\, dt, \tag{1.4.2}$$

when $x \geq y$, and

$$\left(D_{*x}^{\nu} f\right)(y) := \frac{1}{\Gamma(2-\nu)} \int_x^y (y-t)^{1-\nu} f''(t) \, dt, \qquad (1.4.3)$$

when $y \geq x$, where Γ is the gamma function.

We define also the X-valued fractional linear operator

$$(A_0(f))(x, y) := \begin{cases} f'(y) + \left(D_{*y}^{\nu} f\right)(x) \frac{(x-y)^{\nu-1}}{\Gamma(\nu+1)}, & x > y, \\ f'(x) + \left(D_{*x}^{\nu} f\right)(y) \frac{(y-x)^{\nu-1}}{\Gamma(\nu+1)}, & y > x, \\ 0, & x = y. \end{cases} \qquad (1.4.4)$$

By X-valued left fractional Caputo Taylor's formula (see [4]) we get that

$$f(x) - f(y) = f'(y)(x-y) + \frac{1}{\Gamma(\nu)} \int_y^x (x-t)^{\nu-1} D_{*y}^{\nu} f(t) \, dt, \quad \text{for } x > y, \qquad (1.4.5)$$

and

$$f(y) - f(x) = f'(x)(y-x) + \frac{1}{\Gamma(\nu)} \int_x^y (y-t)^{\nu-1} D_{*x}^{\nu} f(t) \, dt, \text{ for } x < y, \qquad (1.4.6)$$

equivalently, it holds

$$f(x) - f(y) = f'(x)(x-y) - \frac{1}{\Gamma(\nu)} \int_x^y (y-t)^{\nu-1} D_{*x}^{\nu} f(t) \, dt, \quad \text{for } x < y. \qquad (1.4.7)$$

We would like to prove that

$$\|f(x) - f(y) - (A_0(f))(x, y)(x-y)\| \leq c \frac{(x-y)^2}{2}, \qquad (1.4.8)$$

for any $x, y \in [0, a]$, $0 < c < 1$.

When $x = y$ the last condition (1.4.8) is trivial.

We assume $x \neq y$. We distinguish the cases:

(1) $x > y$: We observe that

$$\|f(x) - f(y) - (A_0(f))(x, y)(x-y)\| = \qquad (1.4.9)$$

$$\left\| f'(y)(x-y) + \frac{1}{\Gamma(\nu)} \int_y^x (x-t)^{\nu-1} \left(D_{*y}^{\nu} f\right)(t) \, dt - \right.$$

$$\left. \left(f'(y) + \left(D_{*y}^{\nu} f\right)(x) \frac{(x-y)^{\nu-1}}{\Gamma(\nu+1)}\right)(x-y) \right\| =$$

$$\left\| \frac{1}{\Gamma\left(\nu\right)} \int_y^x \left(x-t\right)^{\nu-1} \left(D_{*y}^\nu f\right)(t)\, dt - \left(D_{*y}^\nu f\right)(x) \frac{\left(x-y\right)^\nu}{\Gamma\left(\nu+1\right)} \right\| = \qquad (1.4.10)$$

$$\left\| \frac{1}{\Gamma\left(\nu\right)} \int_y^x \left(x-t\right)^{\nu-1} \left(D_{*y}^\nu f\right)(t)\, dt - \frac{1}{\Gamma\left(\nu\right)} \int_y^x \left(x-t\right)^{\nu-1} \left(D_{*y}^\nu f\right)(x)\, dt \right\| = \qquad (1.4.11)$$

(by [1], p. 426, Theorem 11.43)

$$\frac{1}{\Gamma\left(\nu\right)} \left\| \int_y^x \left(x-t\right)^{\nu-1} \left(\left(D_{*y}^\nu f\right)(t) - \left(D_{*y}^\nu f\right)(x)\right) dt \right\| \le$$

(by [8])

$$\frac{1}{\Gamma\left(\nu\right)} \int_y^x \left(x-t\right)^{\nu-1} \left\| \left(D_{*y}^\nu f\right)(t) - \left(D_{*y}^\nu f\right)(x) \right\| dt =: \left(\xi\right), \qquad (1.4.12)$$

(assume that

$$\left\| \left(D_{*y}^\nu f\right)(t) - \left(D_{*y}^\nu f\right)(x) \right\| \le \lambda_1 \left|t-x\right|^{2-\nu}, \qquad (1.4.13)$$

for any $t, x, y \in [0, a] : x \ge t \ge y$, where $\lambda_1 < \Gamma\left(\nu\right)$, i.e. $\rho_1 := \frac{\lambda_1}{\Gamma\left(\nu\right)} < 1$).
    Therefore

$$\left(\xi\right) \le \frac{\lambda_1}{\Gamma\left(\nu\right)} \int_y^x \left(x-t\right)^{\nu-1} \left(x-t\right)^{2-\nu} dt \qquad (1.4.14)$$

$$= \frac{\lambda_1}{\Gamma\left(\nu\right)} \int_y^x \left(x-t\right) dt = \frac{\lambda_1}{\Gamma\left(\nu\right)} \frac{\left(x-y\right)^2}{2} = \rho_1 \frac{\left(x-y\right)^2}{2}. \qquad (1.4.15)$$

We have proved that

$$\left\| f\left(x\right) - f\left(y\right) - \left(A_0\left(f\right)\right)\left(x, y\right)\left(x-y\right) \right\| \le \rho_1 \frac{\left(x-y\right)^2}{2}, \qquad (1.4.16)$$

where $0 < \rho_1 < 1$, and $x > y$.
    (2) $x < y$: We observe that

$$\left\| f\left(x\right) - f\left(y\right) - \left(A_0\left(f\right)\right)\left(x, y\right)\left(x-y\right) \right\| = \qquad (1.4.17)$$

$$\left\| f'\left(x\right)\left(x-y\right) - \frac{1}{\Gamma\left(\nu\right)} \int_x^y \left(y-t\right)^{\nu-1} D_{*x}^\nu f\left(t\right) dt - \right.$$

$$\left. \left(f'\left(x\right) + \left(D_{*x}^\nu f\right)(y) \frac{\left(y-x\right)^{\nu-1}}{\Gamma\left(\nu+1\right)} \right)\left(x-y\right) \right\| =$$

$$\left\| -\frac{1}{\Gamma(\nu)} \int_x^y (y-t)^{\nu-1} D_{*x}^\nu f(t)\, dt + \left(D_{*x}^\nu f\right)(y) \frac{(y-x)^\nu}{\Gamma(\nu+1)} \right\| = \quad (1.4.18)$$

$$\left\| \frac{1}{\Gamma(\nu)} \int_x^y (y-t)^{\nu-1} D_{*x}^\nu f(t)\, dt - \left(D_{*x}^\nu f\right)(y) \frac{(y-x)^\nu}{\Gamma(\nu+1)} \right\| = \quad (1.4.19)$$

$$\frac{1}{\Gamma(\nu)} \left\| \int_x^y (y-t)^{\nu-1} D_{*x}^\nu f(t)\, dt - \frac{1}{\Gamma(\nu)} \int_x^y (y-t)^{\nu-1} \left(D_{*x}^\nu f\right)(y)\, dt \right\| =$$

$$\frac{1}{\Gamma(\nu)} \left\| \int_x^y (y-t)^{\nu-1} \left(D_{*x}^\nu f(t) - D_{*x}^\nu f(y)\right) dt \right\| \le \quad (1.4.20)$$

$$\frac{1}{\Gamma(\nu)} \int_x^y (y-t)^{\nu-1} \left\| D_{*x}^\nu f(t) - D_{*x}^\nu f(y) \right\| dt$$

(by assumption,

$$\left\| D_{*x}^\nu f(t) - D_{*x}^\nu f(y) \right\| \le \lambda_2 \, |t-y|^{2-\nu}, \quad (1.4.21)$$

for any $t, y, x \in [0, a] : y \ge t \ge x$).

$$\le \frac{1}{\Gamma(\nu)} \int_x^y (y-t)^{\nu-1} \lambda_2 \, |t-y|^{2-\nu}\, dt$$

$$= \frac{\lambda_2}{\Gamma(\nu)} \int_x^y (y-t)^{\nu-1} (y-t)^{2-\nu}\, dt \quad (1.4.22)$$

$$= \frac{\lambda_2}{\Gamma(\nu)} \int_x^y (y-t)\, dt = \frac{\lambda_2}{\Gamma(\nu)} \frac{(x-y)^2}{2}.$$

Assuming also $\rho_2 := \frac{\lambda_2}{\Gamma(\nu)} < 1$ (i.e. $\lambda_2 < \Gamma(\nu)$), we have proved that

$$\| f(x) - f(y) - (A_0(f))(x, y)(x-y) \| \le \rho_2 \frac{(x-y)^2}{2}, \quad \text{for } x < y. \quad (1.4.23)$$

Conclusion: choosing $\lambda := \max(\lambda_1, \lambda_2)$ and $\rho := \frac{\lambda}{\Gamma(\nu)} < 1$, we have proved that

$$\| f(x) - f(y) - (A_0(f))(x, y)(x-y) \| \le \rho \frac{(x-y)^2}{2}, \quad \text{for any } x, y \in [0, a]. \quad (1.4.24)$$

This is a condition needed to solve numerically $f(x) = 0$.

II) Let $n - 1 < \nu < n$, $n \in \mathbb{N} - \{1\}$, i.e. $\lceil \nu \rceil = n$; $x, y \in [0, a]$, $a > 0$, and $f \in C^n([0, a], X)$.

We define the following X-valued right Caputo fractional derivatives (see [3]),

$$D_{x-}^{\nu} f(y) = \frac{(-1)^n}{\Gamma(n-\nu)} \int_y^x (z-y)^{n-\nu-1} f^{(n)}(z) \, dz, \quad \text{for } y \le x, \qquad (1.4.25)$$

and

$$D_{y-}^{\nu} f(x) = \frac{(-1)^n}{\Gamma(n-\nu)} \int_x^y (z-x)^{n-\nu-1} f^{(n)}(z) \, dz, \quad \text{for } x \le y. \qquad (1.4.26)$$

By X-valued right Caputo fractional Taylor's formula (see [3]) we have

$$f(x) - f(y) = \sum_{k=1}^{n-1} \frac{f^{(k)}(y)}{k!} (x-y)^k + \frac{1}{\Gamma(\nu)} \int_x^y (z-x)^{\nu-1} \left(D_{y-}^{\nu} f \right)(z) \, dz,$$
$$(1.4.27)$$

when $x \le y$, and

$$f(y) - f(x) = \sum_{k=1}^{n-1} \frac{f^{(k)}(x)}{k!} (y-x)^k + \frac{1}{\Gamma(\nu)} \int_y^x (z-y)^{\nu-1} \left(D_{x-}^{\nu} f \right)(z) \, dz,$$
$$(1.4.28)$$

when $x \ge y$.

We define also the fractional linear operator

$$(A_0(f))(x, y) := \begin{cases} \sum_{k=1}^{n-1} \frac{f^{(k)}(x)}{k!} (y-x)^k - \left(D_{x-}^{\nu} f \right)(y) \frac{(x-y)^{\nu-1}}{\Gamma(\nu+1)}, & x > y, \\ \sum_{k=1}^{n-1} \frac{f^{(k)}(y)}{k!} (x-y)^k - \left(D_{y-}^{\nu} f \right)(x) \frac{(y-x)^{\nu-1}}{\Gamma(\nu+1)}, & y > x, \\ 0, & x = y. \end{cases}$$
$$(1.4.29)$$

We would like to prove that

$$\| f(x) - f(y) - (A_0(f))(x, y)(x-y) \| \le c \frac{|x-y|^n}{n}, \qquad (1.4.30)$$

for any $x, y \in [0, a]$, $0 < c < 1$.

When $x = y$ the last condition (1.4.30) is trivial.

We assume $x \ne y$. We distinguish the cases:

(1) $x > y$: We observe that

$$\| (f(x) - f(y)) - (A_0(f))(x, y)(x-y) \| = \qquad (1.4.31)$$

$$\| (f(y) - f(x)) - (A_0(f))(x, y)(y-x) \| =$$

$$\left\| \left(\sum_{k=1}^{n-1} \frac{f^{(k)}(x)}{k!} (y-x)^k + \frac{1}{\Gamma(\nu)} \int_y^x (z-y)^{\nu-1} \left(D_{x-}^{\nu} f \right)(z) \, dz \right) - \right.$$

$$\left\| \left(\sum_{k=1}^{n-1} \frac{f^{(k)}(x)}{k!} (y-x)^{k-1} - (D_{x-}^{\nu} f)(y) \frac{(x-y)^{\nu-1}}{\Gamma(\nu+1)} \right)(y-x) \right\| =$$

$$\left\| \frac{1}{\Gamma(\nu)} \int_y^x (z-y)^{\nu-1} (D_{x-}^{\nu} f)(z)\, dz + (D_{x-}^{\nu} f)(y) \frac{(x-y)^{\nu-1}}{\Gamma(\nu+1)} (y-x) \right\| =$$

(1.4.32)

$$\left\| \frac{1}{\Gamma(\nu)} \int_y^x (z-y)^{\nu-1} (D_{x-}^{\nu} f)(z)\, dz - (D_{x-}^{\nu} f)(y) \frac{(x-y)^{\nu}}{\Gamma(\nu+1)} \right\| =$$

$$\frac{1}{\Gamma(\nu)} \left\| \int_y^x (z-y)^{\nu-1} (D_{x-}^{\nu} f)(z)\, dz - \int_y^x (z-y)^{\nu-1} (D_{x-}^{\nu} f)(y)\, dz \right\| =$$

$$\frac{1}{\Gamma(\nu)} \left\| \int_y^x (z-y)^{\nu-1} \left((D_{x-}^{\nu} f)(z) - (D_{x-}^{\nu} f)(y) \right) dz \right\| \leq \qquad (1.4.33)$$

$$\frac{1}{\Gamma(\nu)} \int_y^x (z-y)^{\nu-1} \left\| (D_{x-}^{\nu} f)(z) - (D_{x-}^{\nu} f)(y) \right\| dz$$

(we assume that

$$\left\| (D_{x-}^{\nu} f)(z) - (D_{x-}^{\nu} f)(y) \right\| \leq \lambda_1 |z-y|^{n-\nu}, \qquad (1.4.34)$$

$\lambda_1 > 0$, for all $x, z, y \in [0, a]$, with $x \geq z \geq y$)

$$\leq \frac{\lambda_1}{\Gamma(\nu)} \int_y^x (z-y)^{\nu-1} (z-y)^{n-\nu}\, dz = \qquad (1.4.35)$$

$$= \frac{\lambda_1}{\Gamma(\nu)} \int_y^x (z-y)^{n-1}\, dz = \frac{\lambda_1}{\Gamma(\nu)} \frac{(x-y)^n}{n}$$

(assume $\lambda_1 < \Gamma(\nu)$, i.e. $\rho_1 := \frac{\lambda_1}{\Gamma(\nu)} < 1$)

$$= \rho_1 \frac{(x-y)^n}{n}.$$

We have proved, when $x > y$, that

$$\| f(x) - f(y) - (A_0(f))(x, y)(x-y) \| \leq \rho_1 \frac{(x-y)^n}{n}. \qquad (1.4.36)$$

(2) $y > x$: We observe that

$$\| f(x) - f(y) - (A_0(f))(x, y)(x-y) \| =$$

$$\left\| \left(\sum_{k=1}^{n-1} \frac{f^{(k)}(y)}{k!} (x-y)^k + \frac{1}{\Gamma(\nu)} \int_x^y (z-x)^{\nu-1} \left(D_{y-}^\nu f \right)(z)\, dz \right) - \right.$$

$$\left. \left(\sum_{k=1}^{n-1} \frac{f^{(k)}(y)}{k!} (x-y)^{k-1} - \left(D_{y-}^\nu f \right)(x) \frac{(y-x)^{\nu-1}}{\Gamma(\nu+1)} \right)(x-y) \right\| = \quad (1.4.37)$$

$$\left\| \frac{1}{\Gamma(\nu)} \int_x^y (z-x)^{\nu-1} \left(D_{y-}^\nu f \right)(z)\, dz - \left(D_{y-}^\nu f \right)(x) \frac{(y-x)^\nu}{\Gamma(\nu+1)} \right\| = \quad (1.4.38)$$

$$\left\| \frac{1}{\Gamma(\nu)} \int_x^y (z-x)^{\nu-1} \left(D_{y-}^\nu f \right)(z)\, dz - \frac{1}{\Gamma(\nu)} \int_x^y (z-x)^{\nu-1} \left(D_{y-}^\nu f \right)(x)\, dz \right\| =$$

$$\frac{1}{\Gamma(\nu)} \left\| \int_x^y (z-x)^{\nu-1} \left(\left(D_{y-}^\nu f \right)(z) - \left(D_{y-}^\nu f \right)(x) \right) dz \right\| \leq \quad (1.4.39)$$

$$\frac{1}{\Gamma(\nu)} \int_x^y (z-x)^{\nu-1} \left\| \left(D_{y-}^\nu f \right)(z) - \left(D_{y-}^\nu f \right)(x) \right\| dz$$

(we assume that

$$\left\| \left(D_{y-}^\nu f \right)(z) - \left(D_{y-}^\nu f \right)(x) \right\| \leq \lambda_2 \, |z-x|^{n-\nu}, \quad (1.4.40)$$

$\lambda_2 > 0$, for all $y, z, x \in [0, a]$ with $y \geq z \geq x$)

$$\leq \frac{\lambda_2}{\Gamma(\nu)} \int_x^y (z-x)^{\nu-1} (z-x)^{n-\nu}\, dz = \quad (1.4.41)$$

$$\frac{\lambda_2}{\Gamma(\nu)} \int_x^y (z-x)^{n-1}\, dz = \frac{\lambda_2}{\Gamma(\nu)} \frac{(y-x)^n}{n}.$$

Assume now that $\lambda_2 < \Gamma(\nu)$, that is $\rho_2 := \frac{\lambda_2}{\Gamma(\nu)} < 1$.
    We have proved, for $y > x$, that

$$\| f(x) - f(y) - (A_0(f))(x, y)(x-y) \| \leq \rho_2 \frac{(y-x)^n}{n}. \quad (1.4.42)$$

Set $\lambda := \max(\lambda_1, \lambda_2)$, and

$$0 < \rho := \frac{\lambda}{\Gamma(\nu)} < 1. \quad (1.4.43)$$

Conclusion: We have proved that

$$\|f(x) - f(y) - (A_0(f))(x, y)(x - y)\| \le \rho \frac{|x - y|^n}{n}, \quad \text{for any } x, y \in [0, a].$$
(1.4.44)

In the special case of $1 < \nu < 2$, we obtain that

$$\|f(x) - f(y) - (A_0(f))(x, y)(x - y)\| \le \rho \frac{(x - y)^2}{2},$$
(1.4.45)

for any $x, y \in [0, a], 0 < \rho < 1$.

This is a condition needed to solve numerically $f(x) = 0$.

References

1. C.D. Aliprantis, K.C. Border, *Infinite Dimensional Analysis* (Springer, New York, 2006)
2. S. Amat, S. Busquier, S. Plaza, Chaotic dynamics of a third-order Newton-type method. J. Math. Anal. Appl. **366**(1), 164–174 (2010)
3. G.A. Anastassiou, Strong right fractional calculus for banach space valued functions. Revis. Proyecc. **36**(1), 149–186 (2017)
4. G.A. Anastassiou, A strong fractional calculus theory for banach space valued functions, in *Nonlinear Functional Analysis and Applications* (Korea) (2017). accepted for publication
5. G.A. Anastassiou, I.K. Argyros, Iterative methods and their applications to Banach space valued functions in abstract fractional calculus, in *Progress in Fractional Differentiation and Applications* (2017). accepted
6. I.K. Argyros, A unifying local-semilocal convergence analysis and applications for two-point Newton-like methods in Banach space. J. Math. Anal. Appl. **298**, 374–397 (2004)
7. I.K. Argyros, A. Magréñan, *Iterative Methods and their Dynamics with Applications* (CRC Press, New York, 2017)
8. Bochner integral, *Encyclopedia of Mathematics*, http://www.encyclopediaofmath.org/index. php?title=Bochner_integral&oldid=38659
9. M. Edelstein, On fixed and periodic points under contractive mappings. J. Lond. Math. Soc. **37**, 74–79 (1962)
10. J.A. Ezquerro, J.M. Gutierrez, M.A. Hernandez, N. Romero, M.J. Rubio, The Newton method: from Newton to Kantorovich (Spanish). Gac. R. Soc. Mat. Esp. **13**, 53–76 (2010)
11. L.V. Kantorovich, G.P. Akilov, *Functional Analysis in Normed Spaces* (Pergamon Press, New York, 1982)
12. A. Magréñan, A new tool to study real dynamics: the convergence plane. Appl. Math. Comput. **248**, 215–224 (2014)
13. J. Mikusinski, *The Bochner integral* (Academic Press, New York, 1978)
14. F.A. Potra, V. Pták, *Nondiscrete Induction and Iterative Processes* (Pitman Publ, London, 1984)
15. P.D. Proinov, New general convergence theory for iterative processes and its applications to Newton-Kantorovich type theorems. J. Complex. **26**, 3–42 (2010)
16. G.E. Shilov, *Elementary Functional Analysis* (Dover Publications Inc, New York, 1996)

Chapter 2
Convergence of Iterative Methods in Abstract Fractional Calculus

We present a semi-local convergence analysis for a class of iterative methods under generalized conditions. Some applications are suggested including Banach space valued functions of fractional calculus, where all integrals are of Bochner-type. It follows [6].

2.1 Introduction

Sections 2.1–2.3 are prerequisites for Sect. 2.4.

Let B_1, B_2 stand for Banach spaces and let Ω stand for an open subset of B_1. Let also $U(z, \rho) := \{u \in B_1 : \|u - z\| < \rho\}$ and let $\overline{U}(z, \rho)$ stand for the closure of $U(z, \rho)$.

Many problems in Computational Sciences, Engineering, Mathematical Chemistry, Mathematical Physics, Mathematical Economics and other disciplines can be brought in a form like

$$F(x) = 0 \tag{2.1.1}$$

using Mathematical Modeling [1–18], where $F : \Omega \to B_2$ is a continuous operator. The solution x^* of Eq. (2.1.1) is sought in closed form, but this is attainable only in special cases. That explains why most solution methods for such equations are usually iterative. There is a plethora of iterative methods for solving Eq. (2.1.1). We can divide these methods in two categories.

Explicit Methods [7, 8, 12, 16, 17]: Newton's method

$$x_{n+1} = x_n - F'(x_n)^{-1} F(x_n). \tag{2.1.2}$$

Secant method:

$$x_{n+1} = x_n - [x_{n-1}, x_n; F]^{-1} F(x_n), \tag{2.1.3}$$

© Springer International Publishing AG 2018

G. A. Anastassiou and I. K. Argyros, *Functional Numerical Methods: Applications to Abstract Fractional Calculus*, Studies in Systems, Decision and Control 130, https://doi.org/10.1007/978-3-319-69526-6_2

where $[\cdot, \cdot; F]$ denotes a divided difference of order one on $\Omega \times \Omega$ [8, 16, 17].

Newton-like method:

$$x_{n+1} = x_n - E_n^{-1} F(x_n), \tag{2.1.4}$$

where $E_n = E(F)(x_n)$ and $E : \Omega \rightarrow \mathcal{L}(B_1, B_2)$ the space of bounded linear operators from B_1 into B_2. Other explicit methods can be found in [8, 12, 16, 17] and the references there in.

Implicit Methods [7, 10, 12, 17]:

$$F(x_n) + A_n(x_{n+1} - x_n) = 0 \tag{2.1.5}$$

$$x_{n+1} = x_n - A_n^{-1} F(x_n), \tag{2.1.6}$$

where $A_n = A(x_{n+1}, x_n) = A(F)(x_{n+1}, x_n)$ and $A : \Omega \times \Omega \rightarrow \mathcal{L}(B_1, B_2)$.

There is a plethora on local as well as semi-local convergence results for explicit methods [1–9, 11–17]. However, the research on the convergence of implicit methods has received little attention. Authors, usually consider the fixed point problem

$$P_z(x) = x, \tag{2.1.7}$$

where

$$P_z(x) = x + F(z) + A(x, z)(x - z) \tag{2.1.8}$$

or

$$P_z(x) = z - A(x, z)^{-1} F(z) \tag{2.1.9}$$

for methods (2.1.5) and (2.1.6), respectivelly, where $z \in \Omega$ is given. If P is a contraction operator mapping a closed set into itself, then according to the contraction mapping principle [12, 16, 17], P_z has a fixed point x_z^* which can be found using the method of succesive substitutions or Picard's method [17] defined for each fixed n by

$$y_{k+1,n} = P_{x_n}(y_{k,n}), \quad y_{0,n} = x_n, \quad x_{n+1} = \lim_{k \to +\infty} y_{k,n}. \tag{2.1.10}$$

Let us also consider the analogous explicit methods

$$F(x_n) + A(x_n, x_n)(x_{n+1} - x_n) = 0 \tag{2.1.11}$$

$$x_{n+1} = x_n - A(x_n, x_n)^{-1} F(x_n) \tag{2.1.12}$$

$$F(x_n) + A(x_n, x_{n-1})(x_{n+1} - x_n) = 0 \tag{2.1.13}$$

and

$$x_{n+1} = x_n - A(x_n, x_{n-1})^{-1} F(x_n). \tag{2.1.14}$$

In this chapter in Sect. 2.2, we present the semi-local convergence of method (2.1.5) and method (2.1.6). Section 2.3 contains the semi-local convergence of method (2.1.11), method (2.1.12), method (2.1.13) and method (2.1.14). Some applications to Abstract Fractional Calculus are suggested in Sect. 2.4 on a certain Banach space valued functions, where all the integrals are of Bochner-type [9].

2.2 Semi-local Convergence for Implicit Methods

We present the semi-local convergence analysis of method (2.1.6) using conditions (S):

(s_1) $F : \Omega \subset B_1 \rightarrow B_2$ is continuous and $A(x, y) \in \mathcal{L}(B_1, B_2)$ for each $(x, y) \in \Omega \times \Omega$.

(s_2) There exist $\beta > 0$ and $\Omega_0 \subset B_1$ such that $A(x, y)^{-1} \in \mathcal{L}(B_2, B_1)$ for each $(x, y) \in \Omega_0 \times \Omega_0$ and

$$\left\| A(x, y)^{-1} \right\| \leq \beta^{-1}.$$

Set $\Omega_1 = \Omega \cap \Omega_0$.

(s_3) There exists a continuous and nondecreasing function $\psi : [0, +\infty)^3 \rightarrow [0, +\infty)$ such that for each $x, y \in \Omega_1$

$$\| F(x) - F(y) - A(x, y)(x - y) \| \leq$$

$$\beta \psi (\|x - y\|, \|x - x_0\|, \|y - x_0\|) \|x - y\|.$$

(s_4) For each $x \in \Omega_0$ there exists $y \in \Omega_0$ such that

$$y = x - A(y, x)^{-1} F(x).$$

(s_5) For $x_0 \in \Omega_0$ and $x_1 \in \Omega_0$ satisfying (s_4) there exists $\eta \geq 0$ such that

$$\left\| A(x_1, x_0)^{-1} F(x_0) \right\| \leq \eta.$$

(s_6) Define $q(t) := \psi(\eta, t, t)$ for each $t \in [0, +\infty)$. Equation

$$t(1 - q(t)) - \eta = 0$$

has positive solutions. Denote by s the smallest such solution.

(s_7) $\overline{U}(x_0, s) \subset \Omega$, where

$$s = \frac{\eta}{1 - q_0} \quad \text{and} \quad q_0 = \psi(\eta, s, s).$$

Next, we present the semi-local convergence analysis for method (2.1.6) using the conditions (S) and the preceding notation.

Theorem 2.1 *Assume that the conditions (S) hold. Then, sequence $\{x_n\}$ generated by method (2.1.6) starting at $x_0 \in \Omega$ is well defined in $U(x_0, s)$, remains in $U(x_0, s)$ for each $n = 0, 1, 2, \ldots$ and converges to a solution $x^* \in \overline{U}(x_0, s)$ of equation $F(x) = 0$. Moreover, suppose that there exists a continuous and nondecreasing function $\psi_1 : [0, +\infty)^4 \to [0, +\infty)$ such that for each $x, y, z \in \Omega_1$*

$$\|F(x) - F(y) - A(z, y)(x - y)\| \leq$$

$$\beta \psi_1 (\|x - y\|, \|x - x_0\|, \|y - x_0\|, \|z - x_0\|) \|x - y\|$$

and $q_1 = \psi_1(\eta, s, s, s) < 1$.

Then, x^* is the unique solution of equation $F(x) = 0$ in $\overline{U}(x_0, s)$.

Proof By the definition of s and (s_5), we have $x_1 \in U(x_0, s)$. The proof is based on mathematical induction on k. Suppose that $\|x_k - x_{k-1}\| \leq q_0^{k-1}\eta$ and $\|x_k - x_0\| \leq s$.
We get by (2.1.6), $(s_2) - (s_5)$ in turn that

$$\|x_{k+1} - x_k\| = \|A_k^{-1} F(x_k)\| = \|A_k^{-1}(F(x_k) - F(x_{k-1}) - A_{k-1}(x_k - x_{k-1}))\|$$

$$\leq \|A_k^{-1}\| \|F(x_k) - F(x_{k-1}) - A_{k-1}(x_k - x_{k-1})\| \leq$$

$$\beta^{-1} \beta \psi (\|x_k - x_{k-1}\|, \|x_{k-1} - x_0\|, \|y_k - x_0\|) \|x_k - x_{k-1}\| \leq$$

$$\psi(\eta, s, s) \|x_k - x_{k-1}\| = q_0 \|x_k - x_{k-1}\| \leq q_0^k \|x_1 - x_0\| \leq q_0^k \eta \qquad (2.2.1)$$

and

$$\|x_{k+1} - x_0\| \leq \|x_{k+1} - x_k\| + \ldots + \|x_1 - x_0\|$$

$$\leq q_0^k \eta + \ldots + \eta = \frac{1 - q_0^{k+1}}{1 - q_0} \eta < \frac{\eta}{1 - q_0} = s.$$

The induction is completed. Moreover, we have by (2.2.1) that for $m = 0, 1, 2, \ldots$

$$\|x_{k+m} - x_k\| \leq \frac{1 - q_0^m}{1 - q_0} q_0^k \eta.$$

It follows from the preceding inequation that sequence $\{x_k\}$ is complete in a Banach space B_1 and as such it converges to some $x^* \in \overline{U}(x_0, s)$ (since $\overline{U}(x_0, s)$ is a closed ball). By letting $k \to +\infty$ in (2.2.1) we get $F(x^*) = 0$. To show the uniqueness part, let $x^{**} \in U(x_0, s)$ be a solution of equation $F(x) = 0$. By using (2.1.6) and the hypothesis on ψ_1, we obtain in turn that

$$\|x^{**} - x_{k+1}\| = \|x^{**} - x_k + A_k^{-1} F(x_k) - A_k^{-1} F(x^{**})\| \leq$$

$$\left\| A_k^{-1} \right\| \left\| F\left(x^{**}\right) - F\left(x_k\right) - A_k\left(x^{**} - x_k\right) \right\| \le$$

$$\beta^{-1}\beta\psi_1\left(\left\| x^{**} - x_k \right\|, \left\| x_{k-1} - x_0 \right\|, \left\| x_k - x_0 \right\|, \left\| x^{**} - x_0 \right\|\right)\left\| x^{**} - x_k \right\| \le$$

$$q_1\left\| x^{**} - x_k \right\| \le q_1^{k+1}\left\| x^{**} - x_0 \right\|,$$

so $\lim\limits_{k \to +\infty} x_k = x^{**}$. We have shown that $\lim\limits_{k \to +\infty} x_k = x^*$, so $x^* = x^{**}$. ∎

Remark 2.2 (1) The equation in (s_6) is used to determine the smallness of η. It can be replaced by a stronger condition as follows. Choose $\mu \in (0, 1)$. Denote by s_0 the smallest positive solution of equation $q(t) = \mu$. Notice that if function q is strictly increasing, we can set $s_0 = q^{-1}(\mu)$. Then, we can suppose instead of (s_6):

(s_6') $\eta \le (1 - \mu) s_0$

which is a stronger condition than (s_6).

However, we wanted to leave the equation in (s_6) as uncluttered and as weak as possible.

(2) Condition (s_2) can become part of condition (s_3) by considering

$(s_3)'$ There exists a continuous and nondecreasing function $\varphi : [0, +\infty)^3 \to [0, +\infty)$ such that for each $x, y \in \Omega_1$

$$\left\| A(x, y)^{-1}[F(x) - F(y) - A(x, y)(x, y)] \right\| \le$$

$$\varphi(\|x - y\|, \|x - x_0\|, \|y - x_0\|)\|x - y\|.$$

Notice that

$$\varphi(u_1, u_2, u_3) \le \psi(u_1, u_2, u_3)$$

for each $u_1 \ge 0, u_2 \ge 0$ and $u_3 \ge 0$. Similarly, a function φ_1 can replace ψ_1 for yhe uniqueness of the solution part. These replacements are of Mysovskii-type [7, 12, 16] and influence the weaking of the convergence criterion in (s_6), error bounds and the precision of s.

(3) Suppose that there exist $\beta > 0$, $\beta_1 > 0$ and $L \in \mathcal{L}(B_1, B_2)$ with $L^{-1} \in \mathcal{L}(B_2, B_1)$ such that

$$\left\| L^{-1} \right\| \le \beta^{-1}$$

$$\left\| A(x, y) - L \right\| \le \beta_1$$

and

$$\beta_2 := \beta^{-1}\beta_1 < 1.$$

Then, it follows from the Banach lemma on invertible operators [12], and

$$\left\| L^{-1} \right\| \left\| A(x, y) - L \right\| \le \beta^{-1}\beta_1 = \beta_2 < 1$$

that $A(x, y)^{-1} \in \mathcal{L}(B_2, B_1)$. Let $\beta = \frac{\beta^{-1}}{1-\beta_2}$. Then, under these replacements, condition (s_2) is implied, therefore it can be dropped from the conditions (S).

(4) Clearly method (2.1.5) converges under the conditions (S), since (2.1.6) implies (2.1.5).

(5) We wanted to leave condition (s_4) as uncluttered as possible, since in practice Eqs. (2.1.6) (or (2.1.5)) may be solvable in a way avoiding the already mentioned conditions of the contraction mapping principle. However, in what follows we examine the solvability of method (2.1.5) under a stronger version of the contraction mapping principle using the conditions (V) :

$(v_1) = (s_1)$.

(v_2) There exist functions $w_1 : [0, +\infty)^4 \to [0, +\infty)$, $w_2 : [0, +\infty)^4 \to [0, +\infty)$ continuous and nondecreasing such that for each $x, y, z \in \Omega$

$$\|I + A(x, z) - A(y, z)\| \le w_1 (\|x - y\|, \|x - x_0\|, \|y - x_0\|, \|z - x_0\|)$$

$$\|A(x, z) - A(y, z)\| \le w_2 (\|x - y\|, \|x - x_0\|, \|y - x_0\|, \|z - x_0\|) \|x - y\|$$

and

$$w_1 (0, 0, 0, 0) = w_2 (0, 0, 0, 0) = 0.$$

Set

$$h(t, t, t, t) = \begin{cases} w_1 (2t, t, t, t) + w_2 (2t, t, t, t) (t + \|x_0\|), z \ne x_0 \\ w_1 (2t, t, t, 0) + w_2 (2t, t, t, 0) \|x_0\|, z = x_0. \end{cases}$$

(v_3) There exists $\tau > 0$ satisfying

$$h(t, t, t, t) < 1$$

and

$$h(t, t, 0, t) t + \|F(x_0)\| \le t$$

(v_4) $\overline{U}(x_0, \tau) \subseteq D$.

Theorem 2.3 *Suppose that the conditions (V) are satisfied. Then, Eq. (2.1.5) is uniquely solvable for each $n = 0, 1, 2, \ldots$. Moreover, if $A_n^{-1} \in \mathcal{L}(B_2, B_1)$, the Eq. (2.1.6) is also uniquely solvable for each $n = 0, 1, 2, \ldots$*

Proof The result is an application of the contraction mapping principle. Let $x, y, z \in U(x_0, \tau)$. By the definition of operator P_z, (v_2) and (v_3), we get in turn that

$$\|P_z(x) - P_z(y)\| = \|(I + A(x, z) - A(y, z)) (x - y) - (A(x, z) - A(y, z)) z\|$$

$$\le \|I + A(x, z) - A(y, z)\| \|x - y\| + \|A(x, z) - A(y, z)\| \|z\|$$

$$\le [w_1 (\|x - y\|, \|x - x_0\|, \|y - x_0\|, \|z - x_0\|) +$$

$$w_2 \left(\|x - y\|, \|x - x_0\|, \|y - x_0\|, \|z - x_0\| \right) \left(\|z - x_0\| + \|x_0\| \right) \right] \|x - y\|$$

$$\leq h \left(\tau, \tau, \tau, \tau \right) \|x - y\|$$

and

$$\|P_z (x) - x_0\| \leq \|P_z (x) - P_z (x_0)\| + \|P_z (x_0) - x_0\|$$

$$\leq h \left(\|x - x_0\|, \|x - x_0\|, 0, \|z - x_0\| \right) \|x - x_0\| + \|F (x_0)\|$$

$$\leq h \left(\tau, \tau, 0, \tau \right) \tau + \|F (x_0)\| \leq \tau.$$

■

Remark 2.4 Sections 2.2 and 2.3 have an interest independent of Sect. 2.4. It is worth noticing that the results especially of Theorem 2.1 can apply in Abstract Fractional Calculus as illustrated in Sect. 2.4. By specializing function ψ, we can apply the results of say Theorem 2.1 in the examples suggested in Sect. 2.4. In particular for (2.4.1), we choose $\psi (u_1, u_2, u_3) = \frac{c_1 u_1^p}{(p+1)\beta}$ for $u_1 \geq 0, u_2 \geq 0, u_3 \geq 0$ and c_1, p are given in Sect. 2.4. Similar choices for the other examples of Sect. 2.4. It is also worth noticing that estimate (2.4.2) derived in Sect. 2.4 is of independent interest but not needed in Theorem 2.1.

2.3   Semi-local Convergence for Explicit Methods

A specialization of Theorem 2.1 can be utilized to study the semi-local convergence of the explicit methods given in the introduction of this study. In particular, for the study of method (2.1.12) (and consequently of method (2.1.11)), we use the conditions (S') :

(s_1') $F : \Omega \subset B_1 \to B_2$ is continuous and $A (x, x) \in \mathcal{L} (B_1, B_2)$ for each $x \in \Omega$.

(s_2') There exist $\beta > 0$ and $\Omega_0 \subset B_1$ such that $A (x, x)^{-1} \in \mathcal{L} (B_2, B_1)$ for each $x \in \Omega_0$ and

$$\left\| A (x, x)^{-1} \right\| \leq \beta^{-1}.$$

Set $\Omega_1 = \Omega \cap \Omega_0$.

(s_3') There exist continuous and nondecreasing functions $\psi_0 : [0, +\infty)^3 \to [0, +\infty), \psi_2 : [0, +\infty)^3 \to [0, +\infty)$ with $\psi_0 (0, 0, 0) = \psi_2 (0, 0, 0) = 0$ such that for each $x, y \in \Omega_1$

$$\|F (x) - F (y) - A (y, y) (x - y)\| \leq$$

$$\beta \psi_0 \left(\|x - y\|, \|x - x_0\|, \|y - x_0\| \right) \|x - y\|$$

and

$$\|A(x, y) - A(y, y)\| \leq \beta \psi_2 (\|x - y\|, \|x - x_0\|, \|y - x_0\|).$$

Set $\psi = \psi_0 + \psi_2$.

(s_4') There exist $x_0 \in \Omega_0$ and $\eta \geq 0$ such that $A(x_0, x_0)^{-1} \in \mathcal{L}(B_2, B_1)$ and

$$\|A(x_0, x_0)^{-1} F(x_0)\| \leq \eta.$$

(s_5') = (s_6)
(s_6') = (s_7).

Next, we present the following semi-local convergence analysis of method (2.1.12) using the (S') conditions and the preceding notation.

Proposition 2.5 *Suppose that the conditions (S') are satisfied. Then, sequence $\{x_n\}$ generated by method (2.1.12) starting at $x_0 \in \Omega$ is well defined in $U(x_0, s)$, remains in $U(x_0, s)$ for each $n = 0, 1, 2, \ldots$ and converges to a unique solution $x^* \in \overline{U}(x_0, s)$ of equation $F(x) = 0$.*

Proof We follow the proof of Theorem 2.1 but use instead the analogous estimate

$$\|F(x_k)\| = \|F(x_k) - F(x_{k-1}) - A(x_{k-1}, x_{k-1})(x_k - x_{k-1})\| \leq$$

$$\|F(x_k) - F(x_{k-1}) - A(x_k, x_{k-1})(x_k - x_{k-1})\| +$$

$$\|(A(x_k, x_{k-1}) - A(x_{k-1}, x_{k-1}))(x_k - x_{k-1})\| \leq$$

$$\left[\psi_0 (\|x_k - x_{k-1}\|, \|x_{k-1} - x_0\|, \|x_k - x_0\|) + \right.$$

$$\left. \psi_2 (\|x_k - x_{k-1}\|, \|x_{k-1} - x_0\|, \|x_k - x_0\|) \right] \|x_k - x_{k-1}\| =$$

$$\psi (\|x_k - x_{k-1}\|, \|x_{k-1} - x_0\|, \|x_k - x_0\|) \|x_k - x_{k-1}\|.$$

The rest of the proof is identical to the one in Theorem 2.1 until the uniqueness part for which we have the corresponding estimate

$$\|x^{**} - x_{k+1}\| = \|x^{**} - x_k + A_k^{-1} F(x_k) - A_k^{-1} F(x^{**})\| \leq$$

$$\|A_k^{-1}\| \|F(x^{**}) - F(x_k) - A_k(x^{**} - x_k)\| \leq$$

$$\beta^{-1} \beta \psi_0 (\|x^{**} - x_k\|, \|x_{k-1} - x_0\|, \|x_k - x_0\|) \leq$$

$$q \|x^{**} - x_k\| \leq q^{k+1} \|x^{**} - x_0\|.$$

■

Remark 2.6 Comments similar to the ones given in Sect. 2.2 can follows but for method (2.1.13) and method (2.1.14) instead of method (2.1.5) and method (2.1.6), respectively.

2.4 Applications to Abstract Fractional Calculus

Here we deal with Banach space $(X, \|\cdot\|)$ valued functions f of real domain $[a, b]$. All integrals here are of Bochner-type, see [15]. The derivatives of f are defined similarly to numerical ones, see [18], pp. 83–86 and p. 93.

In this section we apply the earlier numerical methods to X-valued fractional calculus for solving $f(x) = 0$.

Here we would like to establish for $[a, b] \subseteq \mathbb{R}, a < b, f \in C^p([a, b], X), p \in \mathbb{N}$, that

$$\|f(y) - f(x) - A(x, y)(y - x)\| \le c_1 \frac{|x - y|^{p+1}}{p + 1}, \tag{2.4.1}$$

$\forall\, x, y \in [a, b]$, where $c_1 > 0$, and

$$\|A(x, x) - A(y, y)\| \le c_2 |x - y|, \tag{2.4.2}$$

with $c_2 > 0$, $\forall\, x, y \in [a, b]$.

Above A stands for a X-valued differential operator to be defined and presented per case in the next, it will be denoted as $A_+(f)$, $A_-(f)$ in the X-valued fractional cases, and $A_0(f)$ in the X-valued ordinary case.

We examine the following cases:

(I) Here see [4, 5].

Let $x, y \in [a, b]$ such that $x \ge y, \nu > 0, \nu \notin \mathbb{N}$, such that $p = [\nu]$, $[\cdot]$ the integral part, $\alpha = \nu - p$ $(0 < \alpha < 1)$.

Let $f \in C^p([a, b], X)$ and define

$$\left(J_\nu^y f\right)(x) := \frac{1}{\Gamma(\nu)} \int_y^x (x - t)^{\nu-1} f(t)\, dt, \ y \le x \le b, \tag{2.4.3}$$

the X-valued left generalized Riemann-Liouville fractional integral.

Here Γ stands for the gamma function.

Clearly here it holds $\left(J_\nu^y f\right)(y) = 0$. We define $\left(J_\nu^y f\right)(x) = 0$ for $x < y$. By [4] $\left(J_\nu^y f\right)(x)$ is a continuous function in x, for a fixed y.

We define the subspace $C_{y+}^\nu([a, b], X)$ of $C^p([a, b], X)$:

$$C_{y+}^\nu([a, b], X) := \left\{ f \in C^p([a, b], X) : J_{1-\alpha}^y f^{(p)} \in C^1([y, b], X) \right\}. \tag{2.4.4}$$

So let $f \in C_{y+}^\nu([a, b], X)$, we define the **$X$-valued generalized ν−fractional derivative of f** over $[y, b]$ as

$$D_y^\nu f = \left(J_{1-\alpha}^y f^{(p)}\right)',$$ (2.4.5)

that is

$$\left(D_y^\nu f\right)(x) = \frac{1}{\Gamma(1-\alpha)} \frac{d}{dx} \int_y^x (x-t)^{-\alpha} f^{(p)}(t)\, dt,$$ (2.4.6)

which exists for $f \in C_{y+}^\nu ([a, b], X)$, for $a \le y \le x \le b$.

Here we consider $f \in C^p ([a, b], X)$ such that $f \in C_{y+}^\nu ([a, b], X)$, for every $y \in [a, b]$, which means also that $f \in C_{x+}^\nu ([a, b], X)$, for every $x \in [a, b]$ (i.e. exchange roles of x and y), we write that as $f \in C_+^\nu ([a, b], X)$.

That is

$$\left(D_x^\nu f\right)(y) = \frac{1}{\Gamma(1-\alpha)} \frac{d}{dy} \int_x^y (y-t)^{-\alpha} f^{(p)}(t)\, dt$$ (2.4.7)

exists for $f \in C_{x+}^\nu ([a, b], X)$, for $a \le x \le y \le b$.

We mention the following left generalized X-valued fractional Taylor formula ($f \in C_{y+}^\nu ([a, b], X)$, $\nu > 1$), see [5].

It holds

$$f(x) - f(y) = \sum_{k=1}^{p-1} \frac{f^{(k)}(y)}{k!} (x-y)^k + \frac{1}{\Gamma(\nu)} \int_y^x (x-t)^{\nu-1} \left(D_y^\nu f\right)(t)\, dt,$$ (2.4.8)

all $x, y \in [a, b]$ with $x \ge y$.

Similarly for $f \in C_{x+}^\nu ([a, b], X)$ we have

$$f(y) - f(x) = \sum_{k=1}^{p-1} \frac{f^{(k)}(x)}{k!} (y-x)^k + \frac{1}{\Gamma(\nu)} \int_x^y (y-t)^{\nu-1} \left(D_x^\nu f\right)(t)\, dt,$$ (2.4.9)

all $x, y \in [a, b]$ with $y \ge x$.

So here we work with $f \in C^p ([a, b], X)$, such that $f \in C_+^\nu ([a, b], X)$.

We define the X-valued left linear fractional operator

$$
(A_+ (f))(x, y) :=
\begin{cases}
\sum_{k=1}^{p-1} \frac{f^{(k)}(y)}{k!} (x-y)^{k-1} + \left(D_y^\nu f\right)(x) \frac{(x-y)^{\nu-1}}{\Gamma(\nu+1)}, & x > y, \\
\sum_{k=1}^{p-1} \frac{f^{(k)}(x)}{k!} (y-x)^{k-1} + \left(D_x^\nu f\right)(y) \frac{(y-x)^{\nu-1}}{\Gamma(\nu+1)}, & y > x, \\
f^{(p-1)}(x), & x = y.
\end{cases}
$$ (2.4.10)

Notice that (see [13], p. 3)

$$\left\| (A_+ (f))(x, x) - (A_+ (f))(y, y) \right\| = \left\| f^{(p-1)}(x) - f^{(p-1)}(y) \right\|$$ (2.4.11)

$$\le \left\| f^{(p)} \right\|_\infty |x - y|, \ \forall \, x, y \in [a, b],$$

so that condition (2.4.2) is fulfilled.

Next we will prove condition (2.4.1). It is trivially true if $x = y$. So we examine the case of $x \neq y$.

We distinguish the subcases:

(1) $x > y$: We observe that

$$\| f(y) - f(x) - A_+(f)(x, y)(y - x) \| =$$

$$\| f(x) - f(y) - A_+(f)(x, y)(x - y) \| \overset{\text{(by (2.4.8), (2.4.10))}}{=}$$

$$\left\| \sum_{k=1}^{p-1} \frac{f^{(k)}(y)}{k!}(x - y)^k + \frac{1}{\Gamma(\nu)} \int_y^x (x - t)^{\nu-1}(D_y^\nu f)(t)\, dt - \right. \tag{2.4.12}$$

$$\left. \sum_{k=1}^{p-1} \frac{f^{(k)}(y)}{k!}(x - y)^k - (D_y^\nu f)(x)\frac{(x - y)^\nu}{\Gamma(\nu + 1)} \right\| =$$

$$\left\| \frac{1}{\Gamma(\nu)} \int_y^x (x - t)^{\nu-1}(D_y^\nu f)(t)\, dt - \frac{1}{\Gamma(\nu)} \int_y^x (x - t)^{\nu-1}(D_y^\nu f)(x)\, dt \right\|$$

by [1], p. 426, Theorem 11.43

$$= \frac{1}{\Gamma(\nu)} \left\| \int_y^x (x - t)^{\nu-1}\left((D_y^\nu f)(t) - (D_y^\nu f)(x)\right) dt \right\| \leq \tag{2.4.13}$$

(by [9])

$$\frac{1}{\Gamma(\nu)} \int_y^x (x - t)^{\nu-1}\left\| (D_y^\nu f)(t) - (D_y^\nu f)(x) \right\| dt$$

(we assume that

$$\left\| (D_y^\nu f)(t) - (D_y^\nu f)(x) \right\| \leq \lambda_1(y)\,|t - x|^{p+1-\nu}, \tag{2.4.14}$$

for all $x, y, t \in [a, b]$ with $x \geq t \geq y$, with $\lambda_1(y) > 0$ and $\sup_{y \in [a,b]} \lambda_1(y) =: \lambda_1 < \infty$, also it is $0 < p + 1 - \nu < 1$)

$$\leq \frac{\lambda_1}{\Gamma(\nu)} \int_y^x (x - t)^{\nu-1}(x - t)^{p+1-\nu}\, dt = \tag{2.4.15}$$

$$\frac{\lambda_1}{\Gamma(\nu)} \int_y^x (x - t)^p\, dt = \frac{\lambda_1}{\Gamma(\nu)}\frac{(x - y)^{p+1}}{(p + 1)}.$$

We have proved condition (2.4.1)

$$\| f(y) - f(x) - A_+(f)(x, y)(y - x) \| \leq \frac{\lambda_1}{\Gamma(\nu)} \frac{(x - y)^{p+1}}{(p + 1)}, \text{ for } x > y.$$
(2.4.16)

(2) $x < y$: We observe that

$$\| f(y) - f(x) - (A_+(f))(x, y)(y - x) \| \overset{\text{(by (2.4.9), (2.4.10))}}{=}$$

$$\left\| \sum_{k=1}^{p-1} \frac{f^{(k)}(x)}{k!} (y - x)^k + \frac{1}{\Gamma(\nu)} \int_x^y (y - t)^{\nu-1} (D_x^\nu f)(t) \, dt - \right.$$
(2.4.17)

$$\left. \sum_{k=1}^{p-1} \frac{f^{(k)}(x)}{k!} (y - x)^k - (D_x^\nu f)(y) \frac{(y - x)^\nu}{\Gamma(\nu + 1)} \right\| =$$

$$\left\| \frac{1}{\Gamma(\nu)} \int_x^y (y - t)^{\nu-1} (D_x^\nu f)(t) \, dt - (D_x^\nu f)(y) \frac{(y - x)^\nu}{\Gamma(\nu + 1)} \right\| =$$

$$\left\| \frac{1}{\Gamma(\nu)} \int_x^y (y - t)^{\nu-1} (D_x^\nu f)(t) \, dt - \frac{1}{\Gamma(\nu)} \int_x^y (y - t)^{\nu-1} (D_x^\nu f)(y) \, dt \right\| =$$
(2.4.18)

$$\frac{1}{\Gamma(\nu)} \left\| \int_x^y (y - t)^{\nu-1} ((D_x^\nu f)(t) - (D_x^\nu f)(y)) \, dt \right\| \leq$$

$$\frac{1}{\Gamma(\nu)} \int_x^y (y - t)^{\nu-1} \left\| (D_x^\nu f)(t) - (D_x^\nu f)(y) \right\| \, dt$$

(we assume here that

$$\left\| (D_x^\nu f)(t) - (D_x^\nu f)(y) \right\| \leq \lambda_2(x) |t - y|^{p+1-\nu},$$
(2.4.19)

for all $x, y, t \in [a, b]$ with $y \geq t \geq x$, with $\lambda_2(x) > 0$ and $\sup_{x \in [a,b]} \lambda_2(x) =: \lambda_2 < \infty$)

$$\leq \frac{\lambda_2}{\Gamma(\nu)} \int_x^y (y - t)^{\nu-1} (y - t)^{p+1-\nu} \, dt =$$
(2.4.20)

$$\frac{\lambda_2}{\Gamma(\nu)} \int_x^y (y - t)^p \, dt = \frac{\lambda_2}{\Gamma(\nu)} \frac{(y - x)^{p+1}}{(p + 1)}.$$

We have proved that

$$\|f(y) - f(x) - (A_+(f))(x, y)(y - x)\| \le \frac{\lambda_2}{\Gamma(\nu)} \frac{(y - x)^{p+1}}{(p + 1)}, \qquad (2.4.21)$$

for all $x, y \in [a, b]$ such that $y > x$.

Call $\lambda := \max(\lambda_1, \lambda_2)$.

Conclusion We have proved condition (2.4.1), in detail that

$$\|f(y) - f(x) - (A_+(f))(x, y)(y - x)\| \le \frac{\lambda}{\Gamma(\nu)} \frac{|x - y|^{p+1}}{(p + 1)}, \forall x, y \in [a, b].$$
$$(2.4.22)$$

(II) Here see [3] and [5].

Let $x, y \in [a, b]$ such that $x \le y$, $\nu > 0$, $\nu \notin \mathbb{N}$, such that $p = [\nu]$, $\alpha = \nu - p$ $(0 < \alpha < 1)$.

Let $f \in C^p([a, b], X)$ and define

$$\left(J_{y-}^{\nu} f\right)(x) := \frac{1}{\Gamma(\nu)} \int_x^y (z - x)^{\nu-1} f(z) \, dz, \ a \le x \le y, \qquad (2.4.23)$$

the X-valued right generalized Riemann-Liouville fractional integral.

Define the subspace of functions

$$C_{y-}^{\nu}([a, b], X) := \left\{ f \in C^p([a, b], X) : J_{y-}^{1-\alpha} f^{(p)} \in C^1([a, y], X) \right\}. \quad (2.4.24)$$

Define the X-valued right generalized ν-fractional derivative of f over $[a, y]$ as

$$D_{y-}^{\nu} f := (-1)^{p-1} \left(J_{y-}^{1-\alpha} f^{(p)}\right)'. \qquad (2.4.25)$$

Notice that

$$J_{y-}^{1-\alpha} f^{(p)}(x) = \frac{1}{\Gamma(1 - \alpha)} \int_x^y (z - x)^{-\alpha} f^{(p)}(z) \, dz, \qquad (2.4.26)$$

exists for $f \in C_{y-}^{\nu}([a, b], X)$, and

$$\left(D_{y-}^{\nu} f\right)(x) = \frac{(-1)^{p-1}}{\Gamma(1 - \alpha)} \frac{d}{dx} \int_x^y (z - x)^{-\alpha} f^{(p)}(z) \, dz. \qquad (2.4.27)$$

I.e.

$$\left(D_{y-}^{\nu} f\right)(x) = \frac{(-1)^{p-1}}{\Gamma(p - \nu + 1)} \frac{d}{dx} \int_x^y (z - x)^{p-\nu} f^{(p)}(z) \, dz. \qquad (2.4.28)$$

Here we consider $f \in C^p([a, b], X)$ such that $f \in C^{\nu}_{y-}([a, b], X)$, for every $y \in [a, b]$, which means also that $f \in C^{\nu}_{x-}([a, b], X)$, for every $x \in [a, b]$ (i.e. exchange roles of x and y), we write that as $f \in C^{\nu}_{-}([a, b], X)$.

That is

$$\left(D^{\nu}_{x-}f\right)(y) = \frac{(-1)^{p-1}}{\Gamma(p - \nu + 1)} \frac{d}{dy} \int_y^x (z - y)^{p-\nu} f^{(p)}(z) \, dz \qquad (2.4.29)$$

exists for $f \in C^{\nu}_{x-}([a, b], X)$, for $a \leq y \leq x \leq b$.

We mention the following X-valued right generalized fractional Taylor formula $(f \in C^{\nu}_{y-}([a, b], X), \nu > 1)$, see [5].

It holds

$$f(x) - f(y) = \sum_{k=1}^{p-1} \frac{f^{(k)}(y)}{k!} (x - y)^k + \frac{1}{\Gamma(\nu)} \int_x^y (z - x)^{\nu-1} \left(D^{\nu}_{y-}f\right)(z) \, dz,$$

$$(2.4.30)$$

all $x, y \in [a, b]$ with $x \leq y$.

Similarly for $f \in C^{\nu}_{x-}([a, b], X)$ we have

$$f(y) - f(x) = \sum_{k=1}^{p-1} \frac{f^{(k)}(x)}{k!} (y - x)^k + \frac{1}{\Gamma(\nu)} \int_y^x (z - y)^{\nu-1} \left(D^{\nu}_{x-}f\right)(z) \, dz,$$

$$(2.4.31)$$

all $x, y \in [a, b]$ with $x \geq y$.

So here we work with $f \in C^p([a, b], X)$, such that $f \in C^{\nu}_{-}([a, b], X)$.

We define the X-valued right linear fractional operator

$$A_-(f)(x, y) := \begin{cases} \sum_{k=1}^{p-1} \frac{f^{(k)}(x)}{k!} (y - x)^{k-1} - \left(D^{\nu}_{x-}f\right)(y) \frac{(x-y)^{\nu-1}}{\Gamma(\nu+1)}, & x > y, \\ \sum_{k=1}^{p-1} \frac{f^{(k)}(y)}{k!} (x - y)^{k-1} - \left(D^{\nu}_{y-}f\right)(x) \frac{(y-x)^{\nu-1}}{\Gamma(\nu+1)}, & y > x, \\ f^{(p-1)}(x), & x = y. \end{cases}$$

$$(2.4.32)$$

Condition (2.4.2) is fulfilled, the same as in (2.4.11), now for $A_-(f)(x, x)$.

We would like to prove that

$$\|f(x) - f(y) - (A_-(f))(x, y)(x - y)\| \leq c \cdot \frac{|x - y|^{p+1}}{p + 1}, \qquad (2.4.33)$$

for any $x, y \in [a, b]$, where $c > 0$.

When $x = y$ the last condition (2.4.33) is trivial. We assume $x \neq y$.

We distinguish the subcases:

(1) $x > y$: We observe that

$$\|(f(x) - f(y)) - (A_-(f))(x, y)(x - y)\| = \qquad (2.4.34)$$

$$\|(f(y) - f(x)) - (A_-(f))(x, y)(y - x)\| =$$

$$\left\|\left(\sum_{k=1}^{p-1}\frac{f^{(k)}(x)}{k!}(y-x)^k+\frac{1}{\Gamma(\nu)}\int_y^x(z-y)^{\nu-1}\left(D_{x-}^\nu f\right)(z)\,dz\right)-\right.$$

$$\left.\left(\sum_{k=1}^{p-1}\frac{f^{(k)}(x)}{k!}(y-x)^{k-1}-\left(D_{x-}^\nu f\right)(y)\frac{(x-y)^{\nu-1}}{\Gamma(\nu+1)}\right)(y-x)\right\|= \qquad (2.4.35)$$

$$\left\|\frac{1}{\Gamma(\nu)}\int_y^x(z-y)^{\nu-1}\left(D_{x-}^\nu f\right)(z)\,dz+\left(D_{x-}^\nu f\right)(y)\frac{(x-y)^{\nu-1}}{\Gamma(\nu+1)}(y-x)\right\|=$$

$$\left\|\frac{1}{\Gamma(\nu)}\int_y^x(z-y)^{\nu-1}\left(D_{x-}^\nu f\right)(z)\,dz-\left(D_{x-}^\nu f\right)(y)\frac{(x-y)^\nu}{\Gamma(\nu+1)}\right\|=$$

$$\frac{1}{\Gamma(\nu)}\left\|\int_y^x(z-y)^{\nu-1}\left(D_{x-}^\nu f\right)(z)\,dz-\int_y^x(z-y)^{\nu-1}\left(D_{x-}^\nu f\right)(y)\,dz\right\|=$$

$$\frac{1}{\Gamma(\nu)}\left\|\int_y^x(z-y)^{\nu-1}\left(\left(D_{x-}^\nu f\right)(z)-\left(D_{x-}^\nu f\right)(y)\right)dz\right\|\le \qquad (2.4.36)$$

$$\frac{1}{\Gamma(\nu)}\int_y^x(z-y)^{\nu-1}\left\|\left(D_{x-}^\nu f\right)(z)-\left(D_{x-}^\nu f\right)(y)\right\|dz$$

(we assume that

$$\left\|\left(D_{x-}^\nu f\right)(z)-\left(D_{x-}^\nu f\right)(y)\right\|\le\lambda_1|z-y|^{p+1-\nu}, \qquad (2.4.37)$$

$\lambda_1>0$, for all $x,z,y\in[a,b]$ with $x\ge z\ge y$)

$$\le\frac{\lambda_1}{\Gamma(\nu)}\int_y^x(z-y)^{\nu-1}(z-y)^{p+1-\nu}\,dz= \qquad (2.4.38)$$

$$\frac{\lambda_1}{\Gamma(\nu)}\int_y^x(z-y)^p\,dz=\frac{\lambda_1}{\Gamma(\nu)}\frac{(x-y)^{p+1}}{p+1}=\rho_1\frac{(x-y)^{p+1}}{p+1},$$

where $\rho_1:=\frac{\lambda_1}{\Gamma(\nu)}>0$.

We have proved, when $x>y$, that

$$\|f(x)-f(y)-(A_-(f))(x,y)(x-y)\|\le\rho_1\frac{(x-y)^{p+1}}{p+1}. \qquad (2.4.39)$$

(2) $y>x$: We observe that

$$\|f(x)-f(y)-(A_-(f))(x,y)(x-y)\|=$$

$$\left\|\left(\sum_{k=1}^{p-1} \frac{f^{(k)}(y)}{k!}(x-y)^k + \frac{1}{\Gamma(\nu)}\int_x^y (z-x)^{\nu-1}\left(D_{y-}^\nu f\right)(z)\,dz\right)-\right.$$

$$\left.\left(\sum_{k=1}^{p-1} \frac{f^{(k)}(y)}{k!}(x-y)^{k-1} - \left(D_{y-}^\nu f\right)(x)\frac{(y-x)^{\nu-1}}{\Gamma(\nu+1)}\right)(x-y)\right\| = \quad (2.4.40)$$

$$\left\|\frac{1}{\Gamma(\nu)}\int_x^y (z-x)^{\nu-1}\left(D_{y-}^\nu f\right)(z)\,dz - \left(D_{y-}^\nu f\right)(x)\frac{(y-x)^\nu}{\Gamma(\nu+1)}\right\| =$$

$$\left\|\frac{1}{\Gamma(\nu)}\int_x^y (z-x)^{\nu-1}\left(D_{y-}^\nu f\right)(z)\,dz - \frac{1}{\Gamma(\nu)}\int_x^y (z-x)^{\nu-1}\left(D_{y-}^\nu f\right)(x)\,dz\right\| =$$
$$\qquad\qquad\qquad\qquad\qquad\qquad\qquad\qquad\qquad\qquad\qquad\qquad (2.4.41)$$

$$\frac{1}{\Gamma(\nu)}\left\|\int_x^y (z-x)^{\nu-1}\left(\left(D_{y-}^\nu f\right)(z) - \left(D_{y-}^\nu f\right)(x)\right)\,dz\right\| \le \qquad (2.4.42)$$

$$\frac{1}{\Gamma(\nu)}\int_x^y (z-x)^{\nu-1}\left\|\left(D_{y-}^\nu f\right)(z) - \left(D_{y-}^\nu f\right)(x)\right\|\,dz$$

(we assume that

$$\left\|\left(D_{y-}^\nu f\right)(z) - \left(D_{y-}^\nu f\right)(x)\right\| \le \lambda_2\,|z-x|^{p+1-\nu}, \qquad (2.4.43)$$

$\lambda_2 > 0$, for all $y, z, x \in [a, b]$ with $y \ge z \ge x$)

$$\le \frac{\lambda_2}{\Gamma(\nu)}\int_x^y (z-x)^{\nu-1}(z-x)^{p+1-\nu}\,dz = \qquad (2.4.44)$$

$$\frac{\lambda_2}{\Gamma(\nu)}\int_x^y (z-x)^p\,dz = \frac{\lambda_2}{\Gamma(\nu)}\frac{(y-x)^{p+1}}{p+1}.$$

We have proved, for $y > x$, that

$$\|f(x) - f(y) - (A_-(f))(x, y)(x-y)\| \le \rho_2\frac{(y-x)^{p+1}}{p+1}, \qquad (2.4.45)$$

where $\rho_2 := \frac{\lambda_2}{\Gamma(\nu)} > 0$.

Set $\lambda := \max(\lambda_1, \lambda_2)$ and $\rho := \frac{\lambda}{\Gamma(\nu)} > 0$.
Conclusion We have proved (2.4.1) that

$$\|f(x) - f(y) - (A_-(f))(x, y)(x-y)\| \le \rho\frac{|x-y|^{p+1}}{p+1}, \qquad (2.4.46)$$

for any $x, y \in [a, b]$.

(III) Let again $f \in C^p([a, b], X)$, $p \in \mathbb{N}$, $x, y \in [a, b]$.
By vector X-valued Taylor's formula we have, see [3, 4, 18],

$$f(x) - f(y) = \sum_{k=1}^{P} \frac{f^{(k)}(y)}{k!} (x - y)^k + \frac{1}{(p-1)!} \int_y^x (x - t)^{p-1} \left(f^{(p)}(t) - f^{(p)}(y) \right) dt,$$

(2.4.47)

$\forall\, x, y \in [a, b]$.
We define the X-valued function

$$(A_0(f))(x, y) := \begin{cases} \sum_{k=1}^{P} \frac{f^{(k)}(y)}{k!} (x - y)^{k-1}, & x \neq y, \\ f^{(p-1)}(x), & x = y. \end{cases}$$

(2.4.48)

Then it holds, by [13], p. 3,

$$\| (A_0(f))(x, x) - (A_0(f))(y, y) \| = \left\| f^{(p-1)}(x) - f^{(p-1)}(y) \right\|$$

(2.4.49)

$$\leq \left\| f^{(p)} \right\|_\infty |x - y|, \ \forall\, x, y \in [a, b],$$

so that condition (2.4.2) is fulfilled.
Next we observe that

$$\| f(x) - f(y) - (A_0(f))(x, y)(x - y) \| =$$

$$\left\| \sum_{k=1}^{P} \frac{f^{(k)}(y)}{k!} (x - y)^k + \frac{1}{(p-1)!} \int_y^x (x - t)^{p-1} \left(f^{(p)}(t) - f^{(p)}(y) \right) dt \right.$$

(2.4.50)

$$\left. - \sum_{k=1}^{P} \frac{f^{(k)}(y)}{k!} (x - y)^k \right\| =$$

$$\frac{1}{(p-1)!} \left\| \int_y^x (x - t)^{p-1} \left(f^{(p)}(t) - f^{(p)}(y) \right) dt \right\| =: (\xi).$$

(2.4.51)

Here we assume that

$$\left\| f^{(p)}(t) - f^{(p)}(y) \right\| \leq c |t - y|, \ \forall\, t, y \in [a, b], c > 0.$$

(2.4.52)

(1) Subcase of $x > y$: We have that (by [9])

$$(\xi) \leq \frac{1}{(p-1)!} \int_y^x (x - t)^{p-1} \left\| f^{(p)}(t) - f^{(p)}(y) \right\| dt \leq$$

$$\frac{c}{(p-1)!} \int_y^x (x - t)^{p-1} (t - y)^{2-1} dt =$$

(2.4.53)

$$c\frac{\Gamma(p)\,\Gamma(2)}{(p-1)!\,\Gamma(p+2)}\,(x-y)^{p+1} = c\frac{(p-1)!}{(p-1)!\,(p+1)!}\,(x-y)^{p+1}$$

$$= \frac{c\,(x-y)^{p+1}}{(p+1)!}.$$

Hence

$$(\xi) \le c\frac{(x-y)^{p+1}}{(p+1)!},\ x > y. \tag{2.4.54}$$

(2) Subcase of $y > x$.

We have that

$$(\xi) = \frac{1}{(p-1)!}\left\|\int_x^y (t-x)^{p-1}\left(f^{(p)}(y) - f^{(p)}(t)\right)dt\right\| \le \tag{2.4.55}$$

$$\frac{1}{(p-1)!}\int_x^y (t-x)^{p-1}\left\|f^{(p)}(y) - f^{(p)}(t)\right\|dt \le$$

$$\frac{c}{(p-1)!}\int_x^y (t-x)^{p-1}\,(y-t)\,dt =$$

$$\frac{c}{(p-1)!}\int_x^y (y-t)^{2-1}\,(t-x)^{p-1}\,dt = \tag{2.4.56}$$

$$\frac{c}{(p-1)!}\frac{\Gamma(2)\,\Gamma(p)}{\Gamma(p+2)}\,(y-x)^{p+1} = \frac{c}{(p-1)!}\frac{(p-1)!}{(p+1)!}\,(y-x)^{p+1}$$

$$= c\frac{(y-x)^{p+1}}{(p+1)!}.$$

That is

$$(\xi) \le c\frac{(y-x)^{p+1}}{(p+1)!},\ y > x. \tag{2.4.57}$$

Therefore it holds

$$(\xi) \le c\frac{|x-y|^{p+1}}{(p+1)!},\ \text{all } x, y \in [a, b]\ \text{such that } x \ne y. \tag{2.4.58}$$

We have found that

$$\|f(x) - f(y) - (A_0(f))(x, y)(x-y)\| \le c\frac{|x-y|^{p+1}}{(p+1)!},\ c > 0, \tag{2.4.59}$$

for all $x \ne y$.

When $x = y$ inequality (2.4.59) holds trivially, so (2.4.1) it is true for any $x, y \in [a, b]$.

References

1. C.D. Aliprantis, K.C. Border, *Infinite Dimensional Analysis* (Springer, New York, 2006)
2. S. Amat, S. Busquier, S. Plaza, Chaotic dynamics of a third-order Newton-type method. J. Math. Anal. Applic. **366**(1), 164–174 (2010)
3. G.A. Anastassiou, Strong right fractional calculus for banach space valued functions. Rev. Proyecc. **36**(1), 149–186 (2017)
4. G.A. Anastassiou, *A strong Fractional Calculus Theory for Banach Space Valued Functions*, Nonlinear Functional Analysis and Applications (Korea) (accepted for publication, 2017)
5. G.A. Anastassiou, *Strong Mixed and Generalized Fractional Calculus for Banach Space Valued Functions*, Mat. Vesnik (2017)
6. G.A. Anastassiou, I.K. Argyros, *Semi-local Convergence of Iterative Methods and Banach Space Valued Functions in Abstract Fractional Calculus* (submitted, 2017)
7. I.K. Argyros, A unifying local-semilocal convergence analysis and applications for two-point Newton-like methods in Banach space. J. Math. Anal. Appl. **298**, 374–397 (2004)
8. I.K. Argyros, A. Magréñan, *Iterative Methods and their Dynamics with Applications* (CRC Press, New York, 2017)
9. Bochner integral, *Encyclopedia of Mathematics*, http://www.encyclopediaofmath.org/index.php?title=Bochner_integral&oldid=38659
10. M. Edelstein, On fixed and periodic points under contractive mappings. J. Lond. Math. Soc. **37**, 74–79 (1962)
11. J.A. Ezquerro, J.M. Gutierrez, M.A. Hernandez, N. Romero, M.J. Rubio, The Newton method: from Newton to Kantorovich (Spanish). Gac. R. Soc. Mat. Esp. **13**, 53–76 (2010)
12. L.V. Kantorovich, G.P. Akilov, *Functional Analysis in Normed Spaces* (Pergamon Press, New York, 1982)
13. G.E. Ladas, V. Lakshmikantham, *Differential Equations in Abstract Spaces* (Academic Press, New York, 1972)
14. A. Magréñan, A new tool to study real dynamics: the convergence plane. Appl. Math. Comput. **248**, 215–224 (2014)
15. J. Mikusinski, *The Bochner Integral* (Academic Press, New York, 1978)
16. F.A. Potra, V. Ptǎk, *Nondiscrete Induction and Iterative Processes* (Pitman Publ., London, 1984)
17. P.D. Proinov, New general convergence theory for iterative processes and its applications to Newton-Kantorovich type theorems. J. Complex. **26**, 3–42 (2010)
18. G.E. Shilov, *Elementary Functional Analysis* (Dover Publications Inc., New York, 1996)

Chapter 3
Equations for Banach Space Valued Functions in Fractional Vector Calculi

The aim of this chapter is to solve equations on Banach space using iterative methods under generalized conditions. The differentiability of the operator involved is not assumed and its domain is not necessarily convex. Several applications are suggested including Banach space valued functions of abstract fractional calculus, where all integrals are of Bochner-type. It follows [5].

3.1 Introduction

Sections 3.1–3.3 are prerequisites for Sect. 3.4.

Let B_1, B_2 denote Banach spaces and let Ω stand for an open subset of B_1. Let also $U(z, \rho) := \{u \in B_1 : \|u - z\| < \rho\}$ and let $\overline{U}(z, \rho)$ denote the closure of $U(z, \rho)$.

Many problems in Computational Sciences, Engineering, Mathematical Chemistry, Mathematical Physics, Mathematical Economics and other disciplines can be brought in a form like

$$F(x) = 0 \qquad (3.1.1)$$

using Mathematical Modeling [1–17], where $F : \Omega \to B_2$ is a continuous operator. The solution x^* of Eq. (3.1.1) is sought in closed form, but this can be achieved only in special cases. That is why most solution methods for such equations are usually iterative. There is a plethora of iterative methods for solving Eq. (3.1.1). We can divide these methods in two categories.

Explicit Methods: Newton's method [7, 8, 12, 16, 17]

$$x_{n+1} = x_n - F'(x_n)^{-1} F(x_n). \qquad (3.1.2)$$

Secant method:

$$x_{n+1} = x_n - [x_{n-1}, x_n; F]^{-1} F(x_n), \qquad (3.1.3)$$

© Springer International Publishing AG 2018
G. A. Anastassiou and I. K. Argyros, *Functional Numerical Methods: Applications to Abstract Fractional Calculus*, Studies in Systems, Decision and Control 130, https://doi.org/10.1007/978-3-319-69526-6_3

where $[\cdot, \cdot; F]$ denotes a divided difference of order one on $\Omega \times \Omega$ [8, 16, 17].

Newton-like method:

$$x_{n+1} = x_n - E_n^{-1} F(x_n),\qquad (3.1.4)$$

where $E_n = E(F)(x_n)$ and $E : \Omega \to \mathcal{L}(B_1, B_2)$ the space of bounded linear operators from B_1 into B_2. Other explicit methods can be found in [8, 12, 16, 17] and the references there in.

Implicit Methods: [7, 10, 12, 17]:

$$F(x_n) + A_n(x_{n+1} - x_n) = 0 \qquad (3.1.5)$$

$$x_{n+1} = x_n - A_n^{-1} F(x_n),\qquad (3.1.6)$$

where $A_n = A(x_{n+1}, x_n) = A(F)(x_{n+1}, x_n)$ and $A : \Omega \times \Omega \to \mathcal{L}(B_1, B_2)$. We also write $A(F)(x, x) = A(x, x) = A(x)$ for each $x \in \Omega$.

There is a plethora on local as well as semi-local convergence results for explicit methods [1–9, 11–17]. However, the research on the convergence of implicit methods has received little attention. Authors, usually consider the fixed point problem

$$P_z(x) = x,\qquad (3.1.7)$$

where

$$P_z(x) = x + F(z) + A(x, z)(x - z) \qquad (3.1.8)$$

or

$$P_z(x) = z - A(x, z)^{-1} F(z) \qquad (3.1.9)$$

for methods (3.1.5) and (3.1.6), respectively, where $z \in \Omega$ is given. If P is a contraction operator mapping a closed set into itself, then according to the contraction mapping principle [12, 16, 17], P_z has a fixed point x_z^* which can be found using the method of successive substitutions or Picard's method [17] defined for each fixed n by

$$y_{k+1,n} = P_{x_n}(y_{k,n}), \quad y_{0,n} = x_n, \quad x_{n+1} = \lim_{k \to +\infty} y_{k,n}.\qquad (3.1.10)$$

Let us also consider the analogous explicit methods

$$F(x_n) + A(x_n, x_n)(x_{n+1} - x_n) = 0 \qquad (3.1.11)$$

$$x_{n+1} = x_n - A(x_n, x_n)^{-1} F(x_n) \qquad (3.1.12)$$

$$F(x_n) + A(x_n, x_{n-1})(x_{n+1} - x_n) = 0 \qquad (3.1.13)$$

and

$$x_{n+1} = x_n - A(x_n, x_{n-1})^{-1} F(x_n).\qquad (3.1.14)$$

In Sect. 3.2 of this chapter, we present the semi-local convergence of method (3.1.5) and method (3.1.6). Section 3.3 contains the semi-local convergence of method (3.1.11), method (3.1.12), method (3.1.13) and method (3.1.14). Several applications to Abstract Fractional Calculus are suggested in Sect. 3.4 on Banach space valued functions, where all the integrals are of Bochner-type [8, 14].

3.2 Semi-local Convergence for Implicit Methods

We present the semi-local convergence analysis of method (3.1.6) using conditions (S):

(s_1) $F : \Omega \subset B_1 \rightarrow B_2$ is continuous and $A(x, y) \in \mathcal{L}(B_1, B_2)$ for each $(x, y) \in \Omega \times \Omega$.

(s_2) There exist $\beta > 0$ and $\Omega_0 \subset B_1$ such that $A(x, y)^{-1} \in \mathcal{L}(B_2, B_1)$ for each $(x, y) \in \Omega_0 \times \Omega_0$ and

$$\left\| A(x, y)^{-1} \right\| \leq \beta^{-1}.$$

Set $\Omega_1 = \Omega \cap \Omega_0$.

(s_3) There exists a continuous and nondecreasing function $\psi : [0, +\infty)^3 \rightarrow [0, +\infty)$ such that for each $x, y \in \Omega_1$

$$\| F(x) - F(y) - A(x, y)(x - y) \| \leq$$

$$\beta \psi (\|x - y\|, \|x - x_0\|, \|y - x_0\|) \|x - y\|.$$

(s_4) For each $x \in \Omega_0$ there exists $y \in \Omega_0$ such that

$$y = x - A(y, x)^{-1} F(x).$$

(s_5) For $x_0 \in \Omega_0$ and $x_1 \in \Omega_0$ satisfying (s_4) there exists $\eta \geq 0$ such that

$$\left\| A(x_1, x_0)^{-1} F(x_0) \right\| \leq \eta.$$

(s_6) Define $q(t) := \psi(\eta, t, t)$ for each $t \in [0, +\infty)$. Equation

$$t(1 - q(t)) - \eta = 0$$

has positive solutions. Denote by s the smallest such solution.

(s_7) $\overline{U}(x_0, s) \subset \Omega$, where

$$s = \frac{\eta}{1 - q_0} \quad \text{and} \quad q_0 = \psi(\eta, s, s).$$

Next, we present the semi-local convergence analysis for method (3.1.6) using the conditions (S) and the preceding notation.

Theorem 3.1 *Assume that the conditions (S) hold. Then, sequence $\{x_n\}$ generated by method (3.1.6) starting at $x_0 \in \Omega$ is well defined in $U(x_0, s)$, remains in $U(x_0, s)$ for each $n = 0, 1, 2, \ldots$ and converges to a solution $x^* \in \overline{U}(x_0, s)$ of equation $F(x) = 0$. Moreover, suppose that there exists a continuous and nondecreasing function $\psi_1 : [0, +\infty)^4 \to [0, +\infty)$ such that for each $x, y, z \in \Omega_1$*

$$\|F(x) - F(y) - A(z, y)(x - y)\| \le$$

$$\beta \psi_1 (\|x - y\|, \|x - x_0\|, \|y - x_0\|, \|z - x_0\|) \|x - y\|$$

and $q_1 = \psi_1(\eta, s, s, s) < 1$.
  Then, x^* is the unique solution of equation $F(x) = 0$ in $\overline{U}(x_0, s)$.

Proof By the definition of s and (s_5), we have $x_1 \in U(x_0, s)$. The proof is based on mathematical induction on k. Suppose that $\|x_k - x_{k-1}\| \le q_0^{k-1} \eta$ and $\|x_k - x_0\| \le s$.
  We get by (3.1.6), $(s_2) - (s_5)$ in turn that

$$\|x_{k+1} - x_k\| = \left\|A_k^{-1} F(x_k)\right\| = \left\|A_k^{-1}(F(x_k) - F(x_{k-1}) - A_{k-1}(x_k - x_{k-1}))\right\|$$

$$\le \left\|A_k^{-1}\right\| \|F(x_k) - F(x_{k-1}) - A_{k-1}(x_k - x_{k-1})\| \le$$

$$\beta^{-1} \beta \psi (\|x_k - x_{k-1}\|, \|x_{k-1} - x_0\|, \|y_k - x_0\|) \|x_k - x_{k-1}\| \le$$

$$\psi(\eta, s, s) \|x_k - x_{k-1}\| = q_0 \|x_k - x_{k-1}\| \le q_0^k \|x_1 - x_0\| \le q_0^k \eta \qquad (3.2.1)$$

and

$$\|x_{k+1} - x_0\| \le \|x_{k+1} - x_k\| + \ldots + \|x_1 - x_0\|$$

$$\le q_0^k \eta + \ldots + \eta = \frac{1 - q_0^{k+1}}{1 - q_0} \eta < \frac{\eta}{1 - q_0} = s.$$

The induction is completed. Moreover, we have by (3.2.1) that for $m = 0, 1, 2, \ldots$

$$\|x_{k+m} - x_k\| \le \frac{1 - q_0^m}{1 - q_0} q_0^k \eta.$$

It follows from the preceding inequation that sequence $\{x_k\}$ is complete in a Banach space B_1 and as such it converges to some $x^* \in \overline{U}(x_0, s)$ (since $\overline{U}(x_0, s)$ is a closed ball). By letting $k \to +\infty$ in (3.2.1) we get $F(x^*) = 0$. To show the uniqueness part, let $x^{**} \in U(x_0, s)$ be a solution of equation $F(x) = 0$. By using (3.1.6) and the hypothesis on ψ_1, we obtain in turn that

$$\left\|x^{**} - x_{k+1}\right\| = \left\|x^{**} - x_k + A_k^{-1} F(x_k) - A_k^{-1} F(x^{**})\right\| \le$$

$$\left\| A_k^{-1} \right\| \left\| F\left(x^{**}\right) - F\left(x_k\right) - A_k\left(x^{**} - x_k\right) \right\| \le$$

$$\beta^{-1}\beta\psi_1\left(\left\| x^{**} - x_k \right\|, \left\| x_{k-1} - x_0 \right\|, \left\| x_k - x_0 \right\|, \left\| x^{**} - x_0 \right\|\right)\left\| x^{**} - x_k \right\| \le$$

$$q_1\left\| x^{**} - x_k \right\| \le q_1^{k+1}\left\| x^{**} - x_0 \right\|,$$

so $\lim_{k\to+\infty} x_k = x^{**}$. We have shown that $\lim_{k\to+\infty} x_k = x^*$, so $x^* = x^{**}$. ∎

Remark 3.2 (1) The equation in (s_6) is used to determine the smallness of η. It can be replaced by a stronger condition as follows. Choose $\mu \in (0, 1)$. Denote by s_0 the smallest positive solution of equation $q(t) = \mu$. Notice that if function q is strictly increasing, we can set $s_0 = q^{-1}(\mu)$. Then, we can suppose instead of (s_6):

(s_6') $\eta \le (1 - \mu) s_0$

which is a stronger condition than (s_6).

However, we wanted to leave the equation in (s_6) as uncluttered and as weak as possible.

(2) Condition (s_2) can become part of condition (s_3) by considering

$(s_3)'$ There exists a continuous and nondecreasing function $\varphi : [0, +\infty)^3 \to [0, +\infty)$ such that for each $x, y \in \Omega_1$

$$\left\| A(x, y)^{-1}\left[F(x) - F(y) - A(x, y)(x, y)\right] \right\| \le$$

$$\varphi\left(\|x - y\|, \|x - x_0\|, \|y - x_0\|\right)\|x - y\|.$$

Notice that

$$\varphi(u_1, u_2, u_3) \le \psi(u_1, u_2, u_3)$$

for each $u_1 \ge 0$, $u_2 \ge 0$ and $u_3 \ge 0$. Similarly, a function φ_1 can replace ψ_1 for the uniqueness of the solution part. These replacements are of Mysovskii-type [7, 12, 16] and influence the weaking of the convergence criterion in (s_6), error bounds and the precision of s.

(3) Suppose that there exist $\beta > 0$, $\beta_1 > 0$ and $L \in \mathcal{L}(B_1, B_2)$ with $L^{-1} \in \mathcal{L}(B_2, B_1)$ such that

$$\left\| L^{-1} \right\| \le \beta^{-1}$$

$$\left\| A(x, y) - L \right\| \le \beta_1$$

and

$$\beta_2 := \beta^{-1}\beta_1 < 1.$$

Then, it follows from the Banach lemma on invertible operators [12], and

$$\left\| L^{-1} \right\| \left\| A(x, y) - L \right\| \le \beta^{-1}\beta_1 = \beta_2 < 1$$

that $A\,(x,y)^{-1} \in \mathcal{L}\,(B_2, B_1)$. Let $\beta = \frac{\beta^{-1}}{1-\beta_2}$. Then, under these replacements, condition (s_2) is implied, therefore it can be dropped from the conditions (S).

(4) Clearly method (3.1.5) converges under the conditions (S), since (3.1.6) implies (3.1.5).

(5) We wanted to leave condition (s_4) as uncluttered as possible, since in practice Eqs. (3.1.6) or (3.1.5) may be solvable in a way avoiding the already mentioned conditions of the contraction mapping principle. However, in what follows we examine the solvability of method (3.1.5) under a stronger version of the contraction mapping principle using the conditions (V):

$(v_1) = (s_1)$.

(v_2) There exist functions $w_1 : [0, +\infty)^4 \to [0, +\infty)$, $w_2 : [0, +\infty)^4 \to [0, +\infty)$ continuous and nondecreasing such that for each $x, y, z \in \Omega$

$$\|I + A\,(x, z) - A\,(y, z)\| \le w_1\,(\|x - y\|, \|x - x_0\|, \|y - x_0\|, \|z - x_0\|)$$

$$\|A\,(x, z) - A\,(y, z)\| \le w_2\,(\|x - y\|, \|x - x_0\|, \|y - x_0\|, \|z - x_0\|)\,\|x - y\|$$

and

$$w_1\,(0, 0, 0, 0) = w_2\,(0, 0, 0, 0) = 0.$$

Set

$$h\,(t, t, t, t) = \begin{cases} w_1\,(2t, t, t, t) + w_2\,(2t, t, t, t)\,(t + \|x_0\|)\,, z \ne x_0 \\ w_1\,(2t, t, t, 0) + w_2\,(2t, t, t, 0)\,\|x_0\|\,, z = x_0. \end{cases}$$

(v_3) There exists $\tau > 0$ satisfying

$$h\,(t, t, t, t) < 1$$

and

$$h\,(t, t, 0, t)\,t + \|F\,(x_0)\| \le t$$

(v_4) $\overline{U}\,(x_0, \tau) \subseteq D$.

Theorem 3.3 *Suppose that the conditions (V) are satisfied. Then, Eq. (3.1.5) is uniquely solvable for each $n = 0, 1, 2, \dots$. Moreover, if $A_n^{-1} \in \mathcal{L}\,(B_2, B_1)$, the Eq. (3.1.6) is also uniquely solvable for each $n = 0, 1, 2, \dots$*

Proof The result is an application of the contraction mapping principle. Let $x, y, z \in U\,(x_0, \tau)$. By the definition of operator P_z, (v_2) and (v_3), we get in turn that

$$\|P_z\,(x) - P_z\,(y)\| = \|(I + A\,(x, z) - A\,(y, z))\,(x - y) - (A\,(x, z) - A\,(y, z))\,z\|$$

$$\le \|I + A\,(x, z) - A\,(y, z)\|\,\|x - y\| + \|A\,(x, z) - A\,(y, z)\|\,\|z\|$$

$$\le [w_1\,(\|x - y\|, \|x - x_0\|, \|y - x_0\|, \|z - x_0\|) +$$

$$w_2 \left(\|x - y\|, \|x - x_0\|, \|y - x_0\|, \|z - x_0\| \right) \left(\|z - x_0\| + \|x_0\| \right) \right] \|x - y\|$$

$$\leq h(\tau, \tau, \tau, \tau) \|x - y\|$$

and

$$\|P_z(x) - x_0\| \leq \|P_z(x) - P_z(x_0)\| + \|P_z(x_0) - x_0\|$$

$$\leq h(\|x - x_0\|, \|x - x_0\|, 0, \|z - x_0\|) \|x - x_0\| + \|F(x_0)\|$$

$$\leq h(\tau, \tau, 0, \tau)\tau + \|F(x_0)\| \leq \tau.$$

∎

Remark 3.4 Sections 3.2 and 3.3 have an interest independent of Sect. 3.4. It is worth noticing that the results especially of Theorem 3.1 can apply in Abstract Fractional Calculus as illustrated in Sect. 3.4. By specializing function ψ, we can apply the results of say Theorem 3.1 in the examples suggested in Sect. 3.4. In particular for (3.4.8), we choose $\psi(u_1, u_2, u_3) = \frac{cu_1^{p-1}}{\beta p}$ for $u_1 \geq 0$, $u_2 \geq 0$, $u_3 \geq 0$ and c, p are given in Sect. 3.4. Similar choices for the other examples of Sect. 3.4.

3.3 Semi-local Convergence for Explicit Methods

A specialization of Theorem 3.1 can be utilized to study the semi-local convergence of the explicit methods given in the introduction of this study. In particular, for the study of method (3.1.12) (and consequently of method (3.1.11)), we use the conditions (S') :

(s_1') $F : \Omega \subset B_1 \to B_2$ is continuous and $A(x, x) \in \mathcal{L}(B_1, B_2)$ for each $x \in \Omega$.

(s_2') There exist $\beta > 0$ and $\Omega_0 \subset B_1$ such that $A(x, x)^{-1} \in \mathcal{L}(B_2, B_1)$ for each $x \in \Omega_0$ and

$$\left\| A(x, x)^{-1} \right\| \leq \beta^{-1}.$$

Set $\Omega_1 = \Omega \cap \Omega_0$.

(s_3') There exist continuous and nondecreasing functions $\psi_0 : [0, +\infty)^3 \to [0, +\infty)$, $\psi_2 : [0, +\infty)^3 \to [0, +\infty)$ with $\psi_0(0, 0, 0) = \psi_2(0, 0, 0) = 0$ such that for each $x, y \in \Omega_1$

$$\|F(x) - F(y) - A(y, y)(x - y)\| \leq$$

$$\beta\psi_0 \left(\|x - y\|, \|x - x_0\|, \|y - x_0\| \right) \|x - y\|$$

and

$$\|A(x, y) - A(y, y)\| \leq \beta\psi_2 \left(\|x - y\|, \|x - x_0\|, \|y - x_0\| \right).$$

Set $\psi = \psi_0 + \psi_2$.

(s_4') There exist $x_0 \in \Omega_0$ and $\eta \geq 0$ such that $A(x_0, x_0)^{-1} \in \mathcal{L}(B_2, B_1)$ and

$$\left\| A(x_0, x_0)^{-1} F(x_0) \right\| \leq \eta.$$

(s_5') = (s_6)

(s_6') = (s_7).

Next, we present the following semi-local convergence analysis of method (3.1.12) using the (S') conditions and the preceding notation.

Proposition 3.5 *Suppose that the conditions (S') are satisfied. Then, sequence $\{x_n\}$ generated by method (3.1.12) starting at $x_0 \in \Omega$ is well defined in $U(x_0, s)$, remains in $U(x_0, s)$ for each $n = 0, 1, 2, \ldots$ and converges to a unique solution $x^* \in \overline{U}(x_0, s)$ of equation $F(x) = 0$.*

Proof We follow the proof of Theorem 3.1 but use instead the analogous estimate

$$\|F(x_k)\| = \|F(x_k) - F(x_{k-1}) - A(x_{k-1}, x_{k-1})(x_k - x_{k-1})\| \leq$$

$$\|F(x_k) - F(x_{k-1}) - A(x_k, x_{k-1})(x_k - x_{k-1})\| +$$

$$\|(A(x_k, x_{k-1}) - A(x_{k-1}, x_{k-1}))(x_k - x_{k-1})\| \leq$$

$$\left[\psi_0 (\|x_k - x_{k-1}\|, \|x_{k-1} - x_0\|, \|x_k - x_0\|) + \right.$$

$$\psi_2 (\|x_k - x_{k-1}\|, \|x_{k-1} - x_0\|, \|x_k - x_0\|) \Big] \|x_k - x_{k-1}\| =$$

$$\psi (\|x_k - x_{k-1}\|, \|x_{k-1} - x_0\|, \|x_k - x_0\|) \|x_k - x_{k-1}\|.$$

The rest of the proof is identical to the one in Theorem 3.1 until the uniqueness part for which we have the corresponding estimate

$$\left\| x^{**} - x_{k+1} \right\| = \left\| x^{**} - x_k + A_k^{-1} F(x_k) - A_k^{-1} F(x^{**}) \right\| \leq$$

$$\left\| A_k^{-1} \right\| \left\| F(x^{**}) - F(x_k) - A_k (x^{**} - x_k) \right\| \leq$$

$$\beta^{-1} \beta \psi_0 \left(\|x^{**} - x_k\|, \|x_{k-1} - x_0\|, \|x_k - x_0\| \right) \leq$$

$$q \left\| x^{**} - x_k \right\| \leq q^{k+1} \left\| x^{**} - x_0 \right\|. \qquad \blacksquare$$

Remark 3.6 Comments similar to the ones given in Sect. 3.2 can follows but for method (3.1.13) and method (3.1.14) instead of method (3.1.5) and method (3.1.6), respectively.

3.4 Applications to X-valued Fractional and Vector Calculi

Here we deal with Banach space $(X, \|\cdot\|)$ valued functions f of real domain $[a, b]$. All integrals here are of Bochner-type, see [14]. The derivatives of f are defined similarly to numerical ones, see [17], pp. 83–86 and p. 93.

We want to solve numerically

$$f(x) = 0. \tag{3.4.1}$$

(I) Application to X-valued Fractional Calculus

Let $p \in \mathbb{N} - \{1\}$ such that $p - 1 < \nu < p$, where $\nu \notin \mathbb{N}, \nu > 0$, i.e. $\lceil \nu \rceil = p$ ($\lceil \cdot \rceil$ ceiling of the number), $a < b$, $f \in C^p([a, b], X)$.

We define the following X-valued left Caputo fractional derivatives (see [3])

$$\left(D_{*y}^{\nu} f\right)(x) := \frac{1}{\Gamma(p - \nu)} \int_y^x (x - t)^{p-\nu-1} f^{(p)}(t)\, dt, \tag{3.4.2}$$

when $x \geq y$, and

$$\left(D_{*x}^{\nu} f\right)(y) := \frac{1}{\Gamma(p - \nu)} \int_x^y (y - t)^{p-\nu-1} f^{(p)}(t)\, dt, \tag{3.4.3}$$

when $y \geq x$, where Γ is the gamma function.

We define also the X-valued linear operator

$$(A_1(f))(x, y) := \begin{cases} \sum_{k=1}^{p-1} \frac{f^{(k)}(y)}{k!}(x - y)^{k-1} + \left(D_{*y}^{\nu} f\right)(x) \frac{(x-y)^{\nu-1}}{\Gamma(\nu+1)}, & x > y, \\ \sum_{k=1}^{p-1} \frac{f^{(k)}(x)}{k!}(y - x)^{k-1} + \left(D_{*x}^{\nu} f\right)(y) \frac{(y-x)^{\nu-1}}{\Gamma(\nu+1)}, & y > x, \\ f^{(p-1)}(x), & x = y. \end{cases} \tag{3.4.4}$$

By X-valued left fractional Caputo Taylor's formula (see [3]), we get that

$$f(x) - f(y) = \sum_{k=1}^{p-1} \frac{f^{(k)}(y)}{k!}(x - y)^k + \frac{1}{\Gamma(\nu)} \int_y^x (x - t)^{\nu-1} D_{*y}^{\nu} f(t)\, dt, \text{ for } x > y, \tag{3.4.5}$$

and

$$f(y) - f(x) = \sum_{k=1}^{p-1} \frac{f^{(k)}(x)}{k!}(y - x)^k + \frac{1}{\Gamma(\nu)} \int_x^y (y - t)^{\nu-1} D_{*x}^{\nu} f(t)\, dt, \text{ for } x < y. \tag{3.4.6}$$

Immediately, we observe that (by [12], p. 3)

$$\|(A_1(f))(x, x) - (A_1(f))(y, y)\| = \left\| f^{(p-1)}(x) - f^{(p-1)}(y) \right\| \tag{3.4.7}$$

$$\leq \left\| f^{(p)} \right\|_\infty |x - y|, \ \forall \, x, y \in [a, b],$$

We would like to prove that

$$\| f(x) - f(y) - (A_1(f))(x, y)(x - y) \| \leq c \frac{|x - y|^p}{p}, \tag{3.4.8}$$

for any $x, y \in [a, b]$ and some constant $0 < c < 1$.

When $x = y$, the last condition (3.4.8) is trivial.

We assume $x \neq y$. We distinguish the cases:

(1) $x > y$: We observe that

$$\| f(x) - f(y) - (A_1(f))(x, y)(x - y) \| = \tag{3.4.9}$$

$$\left\| \sum_{k=1}^{p-1} \frac{f^{(k)}(y)}{k!}(x - y)^k + \frac{1}{\Gamma(\nu)} \int_y^x (x - t)^{\nu-1} D_{*y}^\nu f(t) \, dt - \right.$$

$$\left. \sum_{k=1}^{p-1} \frac{f^{(k)}(y)}{k!}(x - y)^k - (D_{*y}^\nu f)(x) \frac{(x - y)^\nu}{\Gamma(\nu + 1)} \right\| =$$

$$\left\| \frac{1}{\Gamma(\nu)} \int_y^x (x - t)^{\nu-1}(D_{*y}^\nu f)(t) \, dt - \frac{1}{\Gamma(\nu)} \int_y^x (x - t)^{\nu-1}(D_{*y}^\nu f)(x) \, dt \right\|$$

(by [1], p. 426, Theorem 11.43)

$$= \frac{1}{\Gamma(\nu)} \left\| \int_y^x (x - t)^{\nu-1} \left((D_{*y}^\nu f)(t) - (D_{*y}^\nu f)(x) \right) dt \right\|$$

(by [8])

$$\leq \frac{1}{\Gamma(\nu)} \int_y^x (x - t)^{\nu-1} \left\| (D_{*y}^\nu f)(t) - (D_{*y}^\nu f)(x) \right\| dt \tag{3.4.10}$$

(assume that

$$\left\| (D_{*y}^\nu f)(t) - (D_{*y}^\nu f)(x) \right\| \leq \lambda_1 |t - x|^{p-\nu}, \tag{3.4.11}$$

for any $t, x, y \in [a, b] : x \geq t \geq y$, where $\lambda_1 < \Gamma(\nu)$, i.e. $\rho_1 := \frac{\lambda_1}{\Gamma(\nu)} < 1$)

$$\leq \frac{\lambda_1}{\Gamma(\nu)} \int_y^x (x - t)^{\nu-1}(x - t)^{p-\nu} \, dt = \tag{3.4.12}$$

$$\frac{\lambda_1}{\Gamma(\nu)} \int_y^x (x - t)^{p-1} \, dt = \frac{\lambda_1}{\Gamma(\nu)} \frac{(x - y)^p}{p} = \rho_1 \frac{(x - y)^p}{p}. \tag{3.4.13}$$

We have proved that

$$\|f(x) - f(y) - (A_1(f))(x,y)(x-y)\| \le \rho_1 \frac{(x-y)^p}{p},\qquad(3.4.14)$$

where $0 < \rho_1 < 1$, and $x > y$.

(2) $x < y$: We observe that

$$\|f(x) - f(y) - (A_1(f))(x,y)(x-y)\| = \qquad(3.4.15)$$

$$\|f(y) - f(x) - (A_1(f))(x,y)(y-x)\| =$$

$$\left\| \sum_{k=1}^{p-1} \frac{f^{(k)}(x)}{k!}(y-x)^k + \frac{1}{\Gamma(\nu)} \int_x^y (y-t)^{\nu-1} \left(D_{*x}^\nu f\right)(t)\,dt - \right.$$

$$\left. \sum_{k=1}^{p-1} \frac{f^{(k)}(x)}{k!}(y-x)^k - \left(D_{*x}^\nu f\right)(y)\frac{(y-x)^\nu}{\Gamma(\nu+1)} \right\| =$$

$$\left\| \frac{1}{\Gamma(\nu)} \int_x^y (y-t)^{\nu-1}\left(D_{*x}^\nu f\right)(t)\,dt - \frac{1}{\Gamma(\nu)}\int_x^y (y-t)^{\nu-1}\left(D_{*x}^\nu f\right)(y)\,dt \right\| =$$

$$(3.4.16)$$

$$\frac{1}{\Gamma(\nu)} \left\| \int_x^y (y-t)^{\nu-1}\left(\left(D_{*x}^\nu f\right)(t) - \left(D_{*x}^\nu f\right)(y)\right)dt \right\| \le$$

$$\frac{1}{\Gamma(\nu)}\int_x^y (y-t)^{\nu-1}\left\| \left(D_{*x}^\nu f\right)(t) - \left(D_{*x}^\nu f\right)(y)\right\|\,dt \qquad(3.4.17)$$

(we assume that

$$\left\| \left(D_{*x}^\nu f\right)(t) - \left(D_{*x}^\nu f\right)(y)\right\| \le \lambda_2 \,|t-y|^{p-\nu},\qquad(3.4.18)$$

for any $t, y, x \in [a,b]: y \ge t \ge x$)

$$\le \frac{\lambda_2}{\Gamma(\nu)}\int_x^y (y-t)^{\nu-1}(y-t)^{p-\nu}\,dt =$$

$$\frac{\lambda_2}{\Gamma(\nu)}\int_x^y (y-t)^{p-1}\,dt = \frac{\lambda_2}{\Gamma(\nu)}\frac{(y-x)^p}{p}.\qquad(3.4.19)$$

Assuming also

$$\rho_2 := \frac{\lambda_2}{\Gamma(\nu)} < 1 \qquad(3.4.20)$$

(i.e. $\lambda_2 < \Gamma(\nu)$), we have proved that

$$\|f(x) - f(y) - (A_1(f))(x, y)(x - y)\| \le \rho_2 \frac{(y - x)^p}{p}, \text{ for } x < y. \quad (3.4.21)$$

Conclusion: Choosing $\lambda := \max(\lambda_1, \lambda_2)$ and $\rho := \frac{\lambda}{\Gamma(\nu)} < 1$, we have proved that

$$\|f(x) - f(y) - (A_1(f))(x, y)(x - y)\| \le \rho \frac{|x - y|^p}{p}, \text{ for any } x, y \in [a, b]. \quad (3.4.22)$$

(II) Application from Banach space Mathematical Analysis
In [4], we proved the following general X-valued Taylor's formula:

Theorem 3.7 *Let $p \in \mathbb{N}$ and $f \in C^p([A, B], X)$, where $[A, B] \subset \mathbb{R}$ and $(X, \|\cdot\|)$ is a Banach space. Let $g \in C^1([A, B])$, strictly increasing, such that, $g^{-1} \in C^p$ $([g(A) g(B)])$. Let any $a, b \in [A, B]$. Then*

$$f(b) = f(a) + \sum_{k=1}^{p-1} \frac{(f \circ g^{-1})^{(k)}(g(a))}{k!}(g(b) - g(a))^k + R_p(a, b), \quad (3.4.23)$$

where

$$R_p(a, b) = \frac{1}{(p-1)!}\int_a^b (g(b) - g(s))^{p-1}(f \circ g^{-1})^{(p)}(g(s))g'(s)\,ds \quad (3.4.24)$$

$$= \frac{1}{(p-1)!}\int_{g(a)}^{g(b)} (g(b) - t)^{p-1}(f \circ g^{-1})^{(p)}(t)\,dt.$$

Theorem 3.7 will be applied next for $g(x) = e^x$. One can give similar applications for $g = \sin, \cos, \tan$, etc., over suitable intervals.

Proposition 3.8 *Let $f \in C^p([A, B], X)$, $p \in \mathbb{N}$. Then*

$$f(b) = f(a) + \sum_{k=1}^{p-1} \frac{[(f \circ \ln)^{(k)}(e^a)]}{k!}(e^b - e^a)^k + R_p(a, b), \quad (3.4.25)$$

where

$$R_p(a, b) = \frac{1}{(p-1)!}\int_{e^a}^{e^b}(e^b - t)^{p-1}(f \circ \ln)^{(p)}(t)\,dt$$

$$= \frac{1}{(p-1)!}\int_a^b (e^b - e^a)^{p-1}(f \circ \ln)^{(p)}(e^s)e^s\,ds, \ \forall\, a, b \in [A, B].$$

We will use the following variant.

Theorem 3.9 *Let all as in Theorem 3.7. Then*

$$f(\beta) - f(\alpha) = \sum_{k=1}^{p} \frac{\left(f \circ g^{-1}\right)^{(k)}(g(\alpha))}{k!} (g(\beta) - g(\alpha))^k + R_p^*(\alpha, \beta), \quad (3.4.26)$$

where

$$R_p^*(\alpha, \beta) =$$

$$\frac{1}{(p-1)!} \int_\alpha^\beta (g(\beta) - g(s))^{p-1} \left(\left(f \circ g^{-1}\right)^{(p)}(g(s)) - \left(f \circ g^{-1}\right)^{(p)}(g(\alpha))\right) g'(s) \, ds$$
(3.4.27)

$$= \frac{1}{(p-1)!} \int_{g(\alpha)}^{g(\beta)} (g(\beta) - t)^{p-1} \left(\left(f \circ g^{-1}\right)^{(p)}(t) - \left(f \circ g^{-1}\right)^{(p)}(g(\alpha))\right) dt,$$

$\forall \, \alpha, \beta \in [A, B]$.

Proof Easy.   ∎

Remark 3.10 Call $l = f \circ g^{-1}$. Then $l, l', ..., l^{(p)}$ are continuous from $[g(A), g(B)]$ into $f([A, B])$.

Next we estimate $R_p^*(\alpha, \beta)$: We assume that

$$\left\|\left(f \circ g^{-1}\right)^{(p)}(t) - \left(f \circ g^{-1}\right)^{(p)}(g(\alpha))\right\| \le K |t - g(\alpha)|, \quad (3.4.28)$$

$\forall \, t, g(\alpha) \in [g(A), g(B)]$, where $K > 0$.

We distinguish the cases:

(i) if $g(\beta) > g(\alpha)$, then

$$\left\| R_p^*(\alpha, \beta) \right\| \le$$

$$\frac{1}{(p-1)!} \int_{g(\alpha)}^{g(\beta)} (g(\beta) - t)^{p-1} \left\|\left(f \circ g^{-1}\right)^{(p)}(t) - \left(f \circ g^{-1}\right)^{(p)}(g(\alpha))\right\| dt \le$$

$$\frac{K}{(p-1)!} \int_{g(\alpha)}^{g(\beta)} (g(\beta) - t)^{p-1} (t - g(\alpha))^{2-1} \, dt = \quad (3.4.29)$$

$$\frac{K}{(p-1)!} \frac{\Gamma(p)\,\Gamma(2)}{\Gamma(p+2)} (g(\beta) - g(\alpha))^{p+1} =$$

$$\frac{K}{(p-1)!} \frac{(p-1)!}{(p+1)!} (g(\beta) - g(\alpha))^{p+1} = K \frac{(g(\beta) - g(\alpha))^{p+1}}{(p+1)!}. \quad (3.4.30)$$

We have proved that

$$\left\| R_p^* \left(\alpha, \beta \right) \right\| \le K \frac{\left(g \left(\beta \right) - g \left(\alpha \right) \right)^{p+1}}{\left(p + 1 \right)!}, \tag{3.4.31}$$

when $g \left(\beta \right) > g \left(\alpha \right)$.

(ii) if $g \left(\alpha \right) > g \left(\beta \right)$, then

$$\left\| R_p^* \left(\alpha, \beta \right) \right\| =$$

$$\frac{1}{\left(p - 1 \right)!} \left\| \int_{g(\beta)}^{g(\alpha)} \left(t - g \left(\beta \right) \right)^{p-1} \left(\left(f \circ g^{-1} \right)^{(p)} \left(t \right) - \left(f \circ g^{-1} \right)^{(p)} \left(g \left(\alpha \right) \right) \right) dt \right\| \le$$

$$\frac{1}{\left(p - 1 \right)!} \int_{g(\beta)}^{g(\alpha)} \left(t - g \left(\beta \right) \right)^{p-1} \left\| \left(f \circ g^{-1} \right)^{(p)} \left(t \right) - \left(f \circ g^{-1} \right)^{(p)} \left(g \left(\alpha \right) \right) \right\| dt \le \tag{3.4.32}$$

$$\frac{K}{\left(p - 1 \right)!} \int_{g(\beta)}^{g(\alpha)} \left(g \left(\alpha \right) - t \right)^{2-1} \left(t - g \left(\beta \right) \right)^{p-1} dt =$$

$$\frac{K}{\left(p - 1 \right)!} \frac{\Gamma \left(2 \right) \Gamma \left(p \right)}{\Gamma \left(p + 2 \right)} \left(g \left(\alpha \right) - g \left(\beta \right) \right)^{p+1} = \tag{3.4.33}$$

$$\frac{K}{\left(p - 1 \right)!} \frac{\left(p - 1 \right)!}{\left(p + 1 \right)!} \left(g \left(\alpha \right) - g \left(\beta \right) \right)^{p+1} = K \frac{\left(g \left(\alpha \right) - g \left(\beta \right) \right)^{p+1}}{\left(p + 1 \right)!}. \tag{3.4.34}$$

We have proved that

$$\left\| R_p^* \left(\alpha, \beta \right) \right\| \le K \frac{\left(g \left(\alpha \right) - g \left(\beta \right) \right)^{p+1}}{\left(p + 1 \right)!}, \tag{3.4.35}$$

whenever $g \left(\alpha \right) > g \left(\beta \right)$.

Conclusion: It holds

$$\left\| R_p^* \left(\alpha, \beta \right) \right\| \le K \frac{\left| g \left(\alpha \right) - g \left(\beta \right) \right|^{p+1}}{\left(p + 1 \right)!}, \tag{3.4.36}$$

$\forall \, \alpha, \beta \in [A, B]$.

Both sides of (3.4.36) equal zero when $\alpha = \beta$.

We define the following X-valued linear operator:

$$\left(A_3 \left(f \right) \right) \left(x, y \right) := \begin{cases} \sum_{k=1}^{p} \frac{\left(f \circ g^{-1} \right)^{(k)} \left(g(y) \right)}{k!} \left(g \left(x \right) - g \left(y \right) \right)^{k-1}, & \text{when } g \left(x \right) \ne g \left(y \right), \\ f^{(p-1)} \left(x \right), & x = y, \end{cases} \tag{3.4.37}$$

for any $x, y \in [A, B]$.

Easily, we see that ([12], p. 3)

$$\|(A_3(f))(x, x) - (A_3(f))(y, y)\| = \left\| f^{(p-1)}(x) - f^{(p-1)}(y) \right\| \tag{3.4.38}$$

$$\leq \left\| f^{(p)} \right\|_\infty |x - y|, \forall x, y \in [A, B].$$

Next we observe that (case of $g(x) \neq g(y)$)

$$\| f(x) - f(y) - (A_3(f))(x, y)(g(x) - g(y)) \| =$$

$$\left\| \sum_{k=1}^{p} \frac{\left(f \circ g^{-1} \right)^{(k)}(g(y))}{k!} (g(x) - g(y))^k + R_p^*(y, x) \tag{3.4.39} \right.$$

$$\left. - \left(\sum_{k=1}^{p} \frac{\left(f \circ g^{-1} \right)^{(k)}(g(y))}{k!} (g(x) - g(y))^{k-1}(g(x) - g(y)) \right) \right\| =$$

$$\| R_p^*(y, x) \| \overset{(3.4.36)}{\leq} K \frac{|g(x) - g(y)|^{p+1}}{(p+1)!}, \tag{3.4.40}$$

$\forall x, y \in [A, B] : g(x) \neq g(y)$.

We have proved that

$$\| f(x) - f(y) - (A_3(f))(x, y)(g(x) - g(y)) \| \leq K \frac{|g(x) - g(y)|^{p+1}}{(p+1)!}, \tag{3.4.41}$$

$\forall x, y \in [A, B]$

(the case $x = y$ is trivial).

We apply the above theory as follows:

(II_1) We define

$$(A_{31}(f))(x, y) := \begin{cases} \sum_{k=1}^{p} \frac{(f \circ \ln)^{(k)}(e^y)}{k!} (e^x - e^y)^{k-1}, & x \neq y, \\ f^{(p-1)}(x), & x = y, \end{cases} \tag{3.4.42}$$

for any $x, y \in [A, B]$.

Furthermore it holds

$$\left\| f(x) - f(y) - (A_{31}(f))(x, y)\left(e^x - e^y \right) \right\| \leq K_1 \frac{|e^x - e^y|^{p+1}}{(p+1)!}, \tag{3.4.43}$$

$\forall x, y \in [A, B]$, where we assumed that

$$\left\| (f \circ \ln)^{(p)}(t) - (f \circ \ln)^{(p)}\left(e^y \right) \right\| \leq K_1 \left| t - e^y \right|, \tag{3.4.44}$$

$\forall t, e^y \in \left[e^A, e^B \right]$, $A < B$, with $K_1 > 0$.

(II$_2$) Next let $f \in C^p \left(\left[-\frac{\pi}{2} + \varepsilon, \frac{\pi}{2} - \varepsilon\right]\right)$, $p \in \mathbb{N}$, $\varepsilon > 0$ small.
Here we define that

$$(A_{32}(f))(x, y) := \begin{cases} \sum_{k=1}^{p} \frac{\left(f \circ \sin^{-1}\right)^{(k)}(\sin y)}{k!} (\sin x - \sin y)^{k-1}, & \text{when } x \neq y, \\ f^{(p-1)}(x), & x = y, \end{cases}$$

$$(3.4.45)$$

for any $x, y \in \left[-\frac{\pi}{2} + \varepsilon, \frac{\pi}{2} - \varepsilon\right]$.
We assume that

$$\left\|\left(f \circ \sin^{-1}\right)^{(p)}(t) - \left(f \circ \sin^{-1}\right)^{(p)}(\sin y)\right\| \leq K_2 |t - \sin y|, \qquad (3.4.46)$$

$\forall\, t, \sin y \in \left[\sin\left(-\frac{\pi}{2} + \varepsilon\right), \sin\left(\frac{\pi}{2} - \varepsilon\right)\right]$, where $K_2 > 0$.
It holds

$$\|f(x) - f(y) - (A_{32}(f))(x, y)(\sin x - \sin y)\| \leq K_2 \frac{|\sin x - \sin y|^{p+1}}{(p+1)!},$$

$$(3.4.47)$$

$\forall\, x, y \in \left[-\frac{\pi}{2} + \varepsilon, \frac{\pi}{2} - \varepsilon\right]$.
(II$_3$) Next let $f \in C^p \left([\varepsilon, \pi - \varepsilon]\right)$, $p \in \mathbb{N}$, $\varepsilon > 0$ small.
Here we define

$$(A_{33}(f))(x, y) := \begin{cases} \sum_{k=1}^{p} \frac{\left(f \circ \cos^{-1}\right)^{(k)}(\cos y)}{k!} (\cos x - \cos y)^{k-1}, & \text{when } x \neq y, \\ f^{(p-1)}(x), & x = y, \end{cases}$$

$$(3.4.48)$$

for any $x, y \in [\varepsilon, \pi - \varepsilon]$.
We assume that

$$\left\|\left(f \circ \cos^{-1}\right)^{(p)}(t) - \left(f \circ \cos^{-1}\right)^{(p)}(\cos y)\right\| \leq K_3 |t - \cos y|, \qquad (3.4.49)$$

$\forall\, t, \cos y \in [\cos \varepsilon, \cos(\pi - \varepsilon)]$, where $K_3 > 0$.
It holds

$$\|f(x) - f(y) - (A_{33}(f))(x, y)(\cos x - \cos y)\| \leq K_3 \frac{|\cos x - \cos y|^{p+1}}{(p+1)!},$$

$$(3.4.50)$$

$\forall\, x, y \in [\varepsilon, \pi - \varepsilon]$.
Finally we give:
(II$_4$) Let $f \in C^p \left(\left[-\frac{\pi}{2} + \varepsilon, \frac{\pi}{2} - \varepsilon\right]\right)$, $p \in \mathbb{N}$, $\varepsilon > 0$ small.
We define

$$(A_{34}(f))(x, y) := \begin{cases} \sum_{k=1}^{p} \frac{\left(f \circ \tan^{-1}\right)^{(k)}(\tan y)}{k!} (\tan x - \tan y)^{k-1}, & \text{when } x \neq y, \\ f^{(p-1)}(x), & x = y, \end{cases}$$

$$(3.4.51)$$

for any $x, y \in \left[-\frac{\pi}{2} + \varepsilon, \frac{\pi}{2} - \varepsilon\right]$.

We assume that

$$\left\| \left(f \circ \tan^{-1} \right)^{(p)} (t) - \left(f \circ \tan^{-1} \right)^{(p)} (\tan y) \right\| \le K_4 \left| t - \tan y \right|, \qquad (3.4.52)$$

$\forall\, t, \tan y \in \left[\tan \left(-\frac{\pi}{2} + \varepsilon \right), \tan \left(\frac{\pi}{2} - \varepsilon \right) \right]$, where $K_4 > 0$.
It holds that

$$\left\| f(x) - f(y) - (A_{34}(f))(x, y)(\tan x - \tan y) \right\| \le K_4 \frac{|\tan x - \tan y|^{p+1}}{(p+1)!},$$
$$(3.4.53)$$

$\forall\, x, y \in \left[-\frac{\pi}{2} + \varepsilon, \frac{\pi}{2} - \varepsilon \right].$

References

1. C.D. Aliprantis, K.C. Border, *Infinite Dimensional Analysis* (Springer, New York, 2006)
2. S. Amat, S. Busquier, S. Plaza, Chaotic dynamics of a third-order Newton-type method. J. Math. Anal. Applic. **366**(1), 164–174 (2010)
3. G.A. Anastassiou, A strong fractional calculus theory for banach space valued functions, in *Nonlinear Functional Analysis and Applications* (Korea) (2017). accepted for publication
4. G.A. Anastassiou, Principles of general fractional analysis for Banach space valued functions (2017). submitted for publication
5. G.A. Anastassiou, I.K. Argyros, On the solution of equations and applications on Banach space valued functions and fractional vector calculi (2017). submitted
6. I.K. Argyros, A unifying local-semilocal convergence analysis and applications for two-point Newton-like methods in Banach space. J. Math. Anal. Appl. **298**, 374–397 (2004)
7. I.K. Argyros, A. Magréñan, *Iterative Methods and their Dynamics with Applications* (CRC Press, New York, 2017)
8. Bochner integral, *Encyclopedia of Mathematics*, http://www.encyclopediaofmath.org/index.php?title=Bochner_integral&oldid=38659
9. M. Edelstein, On fixed and periodic points under contractive mappings. J. Lond. Math. Soc. **37**, 74–79 (1962)
10. J.A. Ezquerro, J.M. Gutierrez, M.A. Hernandez, N. Romero, M.J. Rubio, The Newton method: from Newton to Kantorovich (Spanish). Gac. R. Soc. Mat. Esp. **13**, 53–76 (2010)
11. L.V. Kantorovich, G.P. Akilov, *Functional Analysis in Normed Spaces* (Pergamon Press, New York, 1982)
12. G.E. Ladas, V. Lakshmikantham, *Differential Equations in Abstract Spaces* (Academic Press, New York, London, 1972)
13. A. Magréñan, A new tool to study real dynamics: the convergence plane. Appl. Math. Comput. **248**, 215–224 (2014)
14. J. Mikusinski, *The Bochner Integral* (Academic Press, New York, 1978)
15. F.A. Potra, V. Pták, *Nondiscrete Induction and Iterative Processes* (Pitman Publ, London, 1984)
16. P.D. Proinov, New general convergence theory for iterative processes and its applications to Newton-Kantorovich type theorems. J. Complex. **26**, 3–42 (2010)
17. G.E. Shilov, *Elementary Functional Analysis* (Dover Publications Inc, New York, 1996)

Chapter 4
Iterative Methods in Abstract Fractional Calculus

The goal of this chapter is to present a semi-local convergence analysis for some iterative methods under generalized conditions. The operator is only assumed to be continuous and its domain is open. Applications are suggested including Banach space valued functions of fractional calculus, where all integrals are of Bochner-type. It follows [5].

4.1 Introduction

Sections 4.1–4.3 are prerequisites for Sect. 4.4.

Let B_1, B_2 stand for Banach spaces and let Ω stand for an open subset of B_1. Let also $U(z, \rho) := \{u \in B_1 : \|u - z\| < \rho\}$ and let $\overline{U}(z, \rho)$ stand for the closure of $U(z, \rho)$.

Many problems in Computational Sciences, Engineering, Mathematical Chemistry, Mathematical Physics, Mathematical Economics and other disciplines can be brought in a form like

$$F(x) = 0 \tag{4.1.1}$$

using Mathematical Modeling [1–17], where $F : \Omega \to B_2$ is a continuous operator. The solution x^* of Eq. (4.1.1) is sought in closed form. However, this is attainable only in special cases, which explains why most solution methods for such equations are usually iterative. There is a plethora of iterative methods for solving Eq. (4.1.1). We can divide these methods in two categories.

Explicit Methods [7, 8, 12, 16, 17]: Newton's method

$$x_{n+1} = x_n - F'(x_n)^{-1} F(x_n). \tag{4.1.2}$$

Secant method:

$$x_{n+1} = x_n - [x_{n-1}, x_n; F]^{-1} F(x_n), \tag{4.1.3}$$

© Springer International Publishing AG 2018
G. A. Anastassiou and I. K. Argyros, *Functional Numerical Methods: Applications to Abstract Fractional Calculus*, Studies in Systems, Decision and Control 130, https://doi.org/10.1007/978-3-319-69526-6_4

where $[\cdot, \cdot; F]$ denotes a divided difference of order one on $\Omega \times \Omega$ [8, 16, 17].

Newton-like method:

$$x_{n+1} = x_n - E_n^{-1} F(x_n), \tag{4.1.4}$$

where $E_n = E(F)(x_n)$ and $E : \Omega \to \mathcal{L}(B_1, B_2)$ the space of bounded linear operators from B_1 into B_2. Other explicit methods can be found in [8, 12, 16, 17] and the references there in.

Implicit Methods [7, 10, 12, 17]:

$$F(x_n) + A_n(x_{n+1} - x_n) = 0 \tag{4.1.5}$$

$$x_{n+1} = x_n - A_n^{-1} F(x_n), \tag{4.1.6}$$

where $A_n = A(x_{n+1}, x_n) = A(F)(x_{n+1}, x_n)$ and $A : \Omega \times \Omega \to \mathcal{L}(B_1, B_2)$. We denote $A(F)(x, x) = A(x, x) = A(x)$ for each $x \in \Omega$.

There is a plethora on local as well as semi-local convergence results for explicit methods [1–9, 11–17]. However, the research on the convergence of implicit methods has received little attention. Authors, usually consider the fixed point problem

$$P_z(x) = x, \tag{4.1.7}$$

where

$$P_z(x) = x + F(z) + A(x, z)(x - z) \tag{4.1.8}$$

or

$$P_z(x) = z - A(x, z)^{-1} F(z) \tag{4.1.9}$$

for methods (4.1.5) and (4.1.6), respectivelly, where $z \in \Omega$ is given. If P is a contraction operator mapping a closed set into itself, then according to the contraction mapping principle [12, 13, 16, 17], P_z has a fixed point x_z^* which can be found using the method of succesive substitutions or Picard's method [17] defined for each fixed n by

$$y_{k+1,n} = P_{x_n}(y_{k,n}), \quad y_{0,n} = x_n, \ x_{n+1} = \lim_{k \to +\infty} y_{k,n}. \tag{4.1.10}$$

Let us also consider the analogous explicit methods

$$F(x_n) + A(x_n, x_n)(x_{n+1} - x_n) = 0 \tag{4.1.11}$$

$$x_{n+1} = x_n - A(x_n, x_n)^{-1} F(x_n) \tag{4.1.12}$$

$$F(x_n) + A(x_n, x_{n-1})(x_{n+1} - x_n) = 0 \tag{4.1.13}$$

and

$$x_{n+1} = x_n - A(x_n, x_{n-1})^{-1} F(x_n). \tag{4.1.14}$$

In this chapter in Sect. 4.2, we present the semi-local convergence of method (4.1.5) and method (4.1.6). Section 4.3 contains the semi-local convergence of method (4.1.11), method (4.1.12), method (4.1.13) and method (4.1.14). Some applications to Abstract Fractional Calculus are suggested in Sect. 4.4 on a certain Banach space valued functions, where all the integrals are of Bochner-type [8, 14].

4.2 Semi-local Convergence for Implicit Methods

We present the semi-local convergence analysis of method (4.1.6) using conditions (S):

(s_1) $F : \Omega \subset B_1 \rightarrow B_2$ is continuous and $A(x, y) \in \mathcal{L}(B_1, B_2)$ for each $(x, y) \in \Omega \times \Omega$.

(s_2) There exist $\beta > 0$ and $\Omega_0 \subset B_1$ such that $A(x, y)^{-1} \in \mathcal{L}(B_2, B_1)$ for each $(x, y) \in \Omega_0 \times \Omega_0$ and

$$\left\| A(x, y)^{-1} \right\| \leq \beta^{-1}.$$

Set $\Omega_1 = \Omega \cap \Omega_0$.

(s_3) There exists a continuous and nondecreasing function $\psi : [0, +\infty)^3 \rightarrow [0, +\infty)$ such that for each $x, y \in \Omega_1$

$$\| F(x) - F(y) - A(x, y)(x - y) \| \leq$$

$$\beta \psi \left(\| x - y \|, \| x - x_0 \|, \| y - x_0 \| \right) \| x - y \|.$$

(s_4) For each $x \in \Omega_0$ there exists $y \in \Omega_0$ such that

$$y = x - A(y, x)^{-1} F(x).$$

(s_5) For $x_0 \in \Omega_0$ and $x_1 \in \Omega_0$ satisfying (s_4) there exists $\eta \geq 0$ such that

$$\left\| A(x_1, x_0)^{-1} F(x_0) \right\| \leq \eta.$$

(s_6) Define $q(t) := \psi(\eta, t, t)$ for each $t \in [0, +\infty)$. Equation

$$t(1 - q(t)) - \eta = 0$$

has positive solutions. Denote by s the smallest such solution.

(s_7) $\overline{U}(x_0, s) \subset \Omega$, where

$$s = \frac{\eta}{1 - q_0} \quad \text{and} \quad q_0 = \psi(\eta, s, s).$$

Next, we present the semi-local convergence analysis for method (4.1.6) using the conditions (S) and the preceding notation.

Theorem 4.1 *Assume that the conditions (S) hold. Then, sequence $\{x_n\}$ generated by method (4.1.6) starting at $x_0 \in \Omega$ is well defined in $U(x_0, s)$, remains in $U(x_0, s)$ for each $n = 0, 1, 2, ...$ and converges to a solution $x^* \in \overline{U}(x_0, s)$ of equation $F(x) = 0$. Moreover, suppose that there exists a continuous and nondecreasing function $\psi_1 : [0, +\infty)^4 \to [0, +\infty)$ such that for each $x, y, z \in \Omega_1$*

$$\|F(x) - F(y) - A(z, y)(x - y)\| \le$$

$$\beta \psi_1 (\|x - y\|, \|x - x_0\|, \|y - x_0\|, \|z - x_0\|) \|x - y\|$$

and $q_1 = \psi_1(\eta, s, s, s) < 1$.

Then, x^* is the unique solution of equation $F(x) = 0$ in $\overline{U}(x_0, s)$.

Proof By the definition of s and (s_5), we have $x_1 \in U(x_0, s)$. The proof is based on mathematical induction on k. Suppose that $\|x_k - x_{k-1}\| \le q_0^{k-1}\eta$ and $\|x_k - x_0\| \le s$.
We get by (4.1.6), $(s_2) - (s_5)$ in turn that

$$\|x_{k+1} - x_k\| = \left\|A_k^{-1} F(x_k)\right\| = \left\|A_k^{-1}(F(x_k) - F(x_{k-1}) - A_{k-1}(x_k - x_{k-1}))\right\|$$

$$\le \left\|A_k^{-1}\right\| \|F(x_k) - F(x_{k-1}) - A_{k-1}(x_k - x_{k-1})\| \le$$

$$\beta^{-1}\beta\psi(\|x_k - x_{k-1}\|, \|x_{k-1} - x_0\|, \|y_k - x_0\|)\|x_k - x_{k-1}\| \le$$

$$\psi(\eta, s, s)\|x_k - x_{k-1}\| = q_0\|x_k - x_{k-1}\| \le q_0^k\|x_1 - x_0\| \le q_0^k\eta \qquad (4.2.1)$$

and

$$\|x_{k+1} - x_0\| \le \|x_{k+1} - x_k\| + ... + \|x_1 - x_0\|$$

$$\le q_0^k\eta + ... + \eta = \frac{1 - q_0^{k+1}}{1 - q_0}\eta < \frac{\eta}{1 - q_0} = s.$$

The induction is completed. Moreover, we have by (4.2.1) that for $m = 0, 1, 2, ...$

$$\|x_{k+m} - x_k\| \le \frac{1 - q_0^m}{1 - q_0} q_0^k\eta.$$

It follows from the preceding inequation that sequence $\{x_k\}$ is complete in a Banach space B_1 and as such it converges to some $x^* \in \overline{U}(x_0, s)$ (since $\overline{U}(x_0, s)$ is a closed ball). By letting $k \to +\infty$ in (4.2.1) we get $F(x^*) = 0$. To show the uniqueness part, let $x^{**} \in U(x_0, s)$ be a solution of equation $F(x) = 0$. By using (4.1.6) and the hypothesis on ψ_1, we obtain in turn that

$$\left\|x^{**} - x_{k+1}\right\| = \left\|x^{**} - x_k + A_k^{-1} F(x_k) - A_k^{-1} F(x^{**})\right\| \le$$

$$\left\| A_k^{-1} \right\| \left\| F\left(x^{**}\right) - F\left(x_k\right) - A_k\left(x^{**} - x_k\right) \right\| \leq$$

$$\beta^{-1}\beta\psi_1\left(\left\|x^{**} - x_k\right\|, \left\|x_{k-1} - x_0\right\|, \left\|x_k - x_0\right\|, \left\|x^{**} - x_0\right\|\right) \left\|x^{**} - x_k\right\| \leq$$

$$q_1 \left\|x^{**} - x_k\right\| \leq q_1^{k+1} \left\|x^{**} - x_0\right\|,$$

so $\lim\limits_{k \to +\infty} x_k = x^{**}$. We have shown that $\lim\limits_{k \to +\infty} x_k = x^*$, so $x^* = x^{**}$. ∎

Remark 4.2 (1) The equation in (s_6) is used to determine the smallness of η. It can be replaced by a stronger condition as follows. Choose $\mu \in (0, 1)$. Denote by s_0 the smallest positive solution of equation $q(t) = \mu$. Notice that if function q is strictly increasing, we can set $s_0 = q^{-1}(\mu)$. Then, we can suppose instead of (s_6):

$\left(s_6'\right)$ $\eta \leq (1 - \mu) s_0$

which is a stronger condition than (s_6).

However, we wanted to leave the equation in (s_6) as uncluttered and as weak as possible.

(2) Condition (s_2) can become part of condition (s_3) by considering

$(s_3)'$ There exists a continuous and nondecreasing function $\varphi : [0, +\infty)^3 \to [0, +\infty)$ such that for each $x, y \in \Omega_1$

$$\left\| A(x, y)^{-1} [F(x) - F(y) - A(x, y)(x, y)] \right\| \leq$$

$$\varphi(\|x - y\|, \|x - x_0\|, \|y - x_0\|) \|x - y\|.$$

Notice that

$$\varphi(u_1, u_2, u_3) \leq \psi(u_1, u_2, u_3)$$

for each $u_1 \geq 0$, $u_2 \geq 0$ and $u_3 \geq 0$. Similarly, a function φ_1 can replace ψ_1 for the uniqueness of the solution part. These replacements are of Mysovskii-type [7, 12, 16] and influence the weaking of the convergence criterion in (s_6), error bounds and the precision of s.

(3) Suppose that there exist $\beta > 0$, $\beta_1 > 0$ and $L \in \mathcal{L}(B_1, B_2)$ with $L^{-1} \in \mathcal{L}(B_2, B_1)$ such that

$$\left\| L^{-1} \right\| \leq \beta^{-1}$$

$$\left\| A(x, y) - L \right\| \leq \beta_1$$

and

$$\beta_2 := \beta^{-1}\beta_1 < 1.$$

Then, it follows from the Banach lemma on invertible operators [12], and

$$\left\| L^{-1} \right\| \left\| A(x, y) - L \right\| \leq \beta^{-1}\beta_1 = \beta_2 < 1$$

that $A(x, y)^{-1} \in \mathcal{L}(B_2, B_1)$. Let $\beta = \frac{\beta-1}{1-\beta_2}$. Then, under these replacements, condition (s_2) is implied, therefore it can be dropped from the conditions (S).

(4) Clearly method (4.1.5) converges under the conditions (S), since (4.1.6) implies (4.1.5).

(5) We wanted to leave condition (s_4) as uncluttered as possible, since in practice Eqs. (4.1.6) or (4.1.5) may be solvable in a way avoiding the already mentioned conditions of the contraction mapping principle. However, in what follows we examine the solvability of method (4.1.5) under a stronger version of the contraction mapping principle using the conditions (V):

$(v_1) = (s_1)$.

(v_2) There exist functions $w_1 : [0, +\infty)^4 \to [0, +\infty)$, $w_2 : [0, +\infty)^4 \to [0, +\infty)$ continuous and nondecreasing such that for each $x, y, z \in \Omega$

$$\|I + A(x, z) - A(y, z)\| \leq w_1(\|x - y\|, \|x - x_0\|, \|y - x_0\|, \|z - x_0\|)$$

$$\|A(x, z) - A(y, z)\| \leq w_2(\|x - y\|, \|x - x_0\|, \|y - x_0\|, \|z - x_0\|) \|x - y\|$$

and

$$w_1(0, 0, 0, 0) = w_2(0, 0, 0, 0) = 0.$$

Set

$$h(t, t, t, t) = \begin{cases} w_1(2t, t, t, t) + w_2(2t, t, t, t)(t + \|x_0\|), & z \neq x_0 \\ w_1(2t, t, t, 0) + w_2(2t, t, t, 0)\|x_0\|, & z = x_0. \end{cases}$$

(v_3) There exists $\tau > 0$ satisfying

$$h(t, t, t, t) < 1$$

and

$$h(t, t, 0, t)t + \|F(x_0)\| \leq t$$

(v_4) $\overline{U}(x_0, \tau) \subseteq D$.

Theorem 4.3 *Suppose that the conditions* (V) *are satisfied. Then, Eq. (4.1.5) is uniquely solvable for each* $n = 0, 1, 2, \ldots$ *Moreover, if* $A_n^{-1} \in \mathcal{L}(B_2, B_1)$, *the Eq. (4.1.6) is also uniquely solvable for each* $n = 0, 1, 2, \ldots$

Proof The result is an application of the contraction mapping principle. Let $x, y, z \in U(x_0, \tau)$. By the definition of operator P_z, (v_2) and (v_3), we get in turn that

$$\|P_z(x) - P_z(y)\| = \|(I + A(x, z) - A(y, z))(x - y) - (A(x, z) - A(y, z))z\|$$

$$\leq \|I + A(x, z) - A(y, z)\| \|x - y\| + \|A(x, z) - A(y, z)\| \|z\|$$

$$\leq [w_1(\|x - y\|, \|x - x_0\|, \|y - x_0\|, \|z - x_0\|) +$$

$$w_2 \left(\|x - y\|, \|x - x_0\|, \|y - x_0\|, \|z - x_0\| \right) \left(\|z - x_0\| + \|x_0\| \right) \right] \|x - y\|$$

$$\leq h \left(\tau, \tau, \tau, \tau \right) \|x - y\|$$

and

$$\| P_z(x) - x_0 \| \leq \| P_z(x) - P_z(x_0) \| + \| P_z(x_0) - x_0 \|$$

$$\leq h \left(\|x - x_0\|, \|x - x_0\|, 0, \|z - x_0\| \right) \|x - x_0\| + \| F(x_0) \|$$

$$\leq h \left(\tau, \tau, 0, \tau \right) \tau + \| F(x_0) \| \leq \tau.$$

∎

Remark 4.4 Sections 4.2 and 4.3 have an interest independent of Sect. 4.4. It is worth noticing that the results especially of Theorem 4.1 can apply in Abstract Fractional Calculus as illustrated in Sect. 4.4. By specializing function ψ, we can apply the results of say Theorem 4.1 in the examples suggested in Sect. 4.4. In particular for (4.4.21), we choose $\psi(u_1, u_2, u_3) = \frac{\lambda u_1^{(n+1)\alpha}}{\beta \Gamma((n+1)\alpha)((n+1)\alpha+1)}$ for $u_1 \geq 0, u_2 \geq 0, u_3 \geq 0$ and λ, α are given in Sect. 4.4. Similar choices for the other examples of Sect. 4.4. It is also worth noticing that estimate (4.4.2) derived in Sect. 4.4 is of independent interest but not needed in Theorem 4.1.

4.3 Semi-local Convergence for Explicit Methods

A specialization of Theorem 4.1 can be utilized to study the semi-local convergence of the explicit methods given in the introduction of this study. In particular, for the study of method (4.1.12) (and consequently of method (4.1.11)), we use the conditions (S'):

(s_1') $F : \Omega \subset B_1 \to B_2$ is continuous and $A(x, x) \in \mathcal{L}(B_1, B_2)$ for each $x \in \Omega$.

(s_2') There exist $\beta > 0$ and $\Omega_0 \subset B_1$ such that $A(x, x)^{-1} \in \mathcal{L}(B_2, B_1)$ for each $x \in \Omega_0$ and

$$\left\| A(x, x)^{-1} \right\| \leq \beta^{-1}.$$

Set $\Omega_1 = \Omega \cap \Omega_0$.

(s_3') There exist continuous and nondecreasing functions $\psi_0 : [0, +\infty)^3 \to [0, +\infty)$, $\psi_2 : [0, +\infty)^3 \to [0, +\infty)$ with $\psi_0(0, 0, 0) = \psi_2(0, 0, 0) = 0$ such that for each $x, y \in \Omega_1$

$$\| F(x) - F(y) - A(y, y)(x - y) \| \leq$$

$$\beta \psi_0 \left(\|x - y\|, \|x - x_0\|, \|y - x_0\| \right) \|x - y\|$$

and

$$\|A(x, y) - A(y, y)\| \leq \beta\psi_2(\|x - y\|, \|x - x_0\|, \|y - x_0\|).$$

Set $\psi = \psi_0 + \psi_2$.

(s_4') There exist $x_0 \in \Omega_0$ and $\eta \geq 0$ such that $A(x_0, x_0)^{-1} \in \mathcal{L}(B_2, B_1)$ and

$$\|A(x_0, x_0)^{-1} F(x_0)\| \leq \eta.$$

$(s_5') = (s_6)$
$(s_6') = (s_7)$.

Next, we present the following semi-local convergence analysis of method (4.1.12) using the (S') conditions and the preceding notation.

Proposition 4.5 *Suppose that the conditions (S') are satisfied. Then, sequence $\{x_n\}$ generated by method (4.1.12) starting at $x_0 \in \Omega$ is well defined in $U(x_0, s)$, remains in $U(x_0, s)$ for each $n = 0, 1, 2, \ldots$ and converges to a unique solution $x^* \in \overline{U}(x_0, s)$ of equation $F(x) = 0$.*

Proof We follow the proof of Theorem 4.1 but use instead the analogous estimate

$$\|F(x_k)\| = \|F(x_k) - F(x_{k-1}) - A(x_{k-1}, x_{k-1})(x_k - x_{k-1})\| \leq$$

$$\|F(x_k) - F(x_{k-1}) - A(x_k, x_{k-1})(x_k - x_{k-1})\| +$$

$$\|(A(x_k, x_{k-1}) - A(x_{k-1}, x_{k-1}))(x_k - x_{k-1})\| \leq$$

$$\left[\psi_0(\|x_k - x_{k-1}\|, \|x_{k-1} - x_0\|, \|x_k - x_0\|) +\right.$$

$$\psi_2(\|x_k - x_{k-1}\|, \|x_{k-1} - x_0\|, \|x_k - x_0\|)\Big] \|x_k - x_{k-1}\| =$$

$$\psi(\|x_k - x_{k-1}\|, \|x_{k-1} - x_0\|, \|x_k - x_0\|) \|x_k - x_{k-1}\|.$$

The rest of the proof is identical to the one in Theorem 4.1 until the uniqueness part for which we have the corresponding estimate

$$\|x^{**} - x_{k+1}\| = \|x^{**} - x_k + A_k^{-1} F(x_k) - A_k^{-1} F(x^{**})\| \leq$$

$$\|A_k^{-1}\| \|F(x^{**}) - F(x_k) - A_k(x^{**} - x_k)\| \leq$$

$$\beta^{-1}\beta\psi_0(\|x^{**} - x_k\|, \|x_{k-1} - x_0\|, \|x_k - x_0\|) \leq$$

$$q\|x^{**} - x_k\| \leq q^{k+1}\|x^{**} - x_0\|. \qquad \blacksquare$$

Remark 4.6 Comments similar to the ones given in Sect. 4.2 can follows but for method (4.1.13) and method (4.1.14) instead of method (4.1.5) and method (4.1.6), respectively.

4.4 Applications to X-valued Fractional Calculus

Here we deal with Banach space $(X, \|\cdot\|)$ valued functions f of real domain $[a, b]$. All integrals are of Bochner-type, see [14]. The derivatives of f are defined similarly to numerical ones, see [17], pp. 83–86 and p. 93.

Let $f : [a, b] \to X$ such that $f^{(m)} \in L_\infty ([a, b], X)$, the X-valued left Caputo fractional derivative of order $\alpha \notin \mathbb{N}$, $\alpha > 0$, $m = \lceil \alpha \rceil$ ($\lceil \cdot \rceil$ ceiling) is defined as follows (see [3]):

$$\left(D_a^\alpha f\right)(x) = \frac{1}{\Gamma(m - \alpha)} \int_a^x (x - t)^{m - \alpha - 1} f^{(m)}(t)\, dt, \qquad (4.4.1)$$

where Γ is the gamma function, $\forall\, x \in [a, b]$.

We observe that

$$\left\| \left(D_a^\alpha f\right)(x) \right\| \le \frac{1}{\Gamma(m - \alpha)} \int_a^x (x - t)^{m - \alpha - 1} \left\| f^{(m)}(t) \right\| dt$$

$$\le \frac{\left\| f^{(m)} \right\|_\infty}{\Gamma(m - \alpha)} \left(\int_a^x (x - t)^{m - \alpha - 1}\, dt \right) = \frac{\left\| f^{(m)} \right\|_\infty}{\Gamma(m - \alpha)} \frac{(x - a)^{m - \alpha}}{(m - \alpha)}$$

$$= \frac{\left\| f^{(m)} \right\|_\infty}{\Gamma(m - \alpha + 1)} (x - a)^{m - \alpha}. \qquad (4.4.2)$$

We have proved that

$$\left\| \left(D_a^\alpha f\right)(x) \right\| \le \frac{\left\| f^{(m)} \right\|_\infty}{\Gamma(m - \alpha + 1)} (x - a)^{m - \alpha} \le \frac{\left\| f^{(m)} \right\|_\infty}{\Gamma(m - \alpha + 1)} (b - a)^{m - \alpha}. \qquad (4.4.3)$$

Clearly then $\left(D_a^\alpha f\right)(a) = 0$.

Let $n \in \mathbb{N}$ we denote $D_a^{n\alpha} = D_a^\alpha D_a^\alpha ... D_a^\alpha$ (n-times).

Let us assume now that

$$f \in C^1([a, b], X), D_a^{k\alpha} f \in C^1([a, b], X), \ k = 1, ..., n;$$

$$D_a^{(n+1)\alpha} f \in C([a, b], X), n \in \mathbb{N}, 0 < \alpha \le 1. \qquad (4.4.4)$$

By [4], we have

$$f(x) = \sum_{i=0}^n \frac{(x - a)^{i\alpha}}{\Gamma(i\alpha + 1)} \left(D_a^{i\alpha} f\right)(a) + \qquad (4.4.5)$$

$$\frac{1}{\Gamma((n+1)\alpha)} \int_a^x (x - t)^{(n+1)\alpha - 1} \left(D_a^{(n+1)\alpha} f\right)(t)\, dt, \ \forall\, x \in [a, b].$$

Under our assumption and conclusion, see (4.4.4), Taylor's formula (4.4.5) becomes

$$f(x) - f(a) = \sum_{i=2}^{n} \frac{(x-a)^{i\alpha}}{\Gamma(i\alpha+1)} \left(D_a^{i\alpha} f\right)(a) +$$

$$\frac{1}{\Gamma((n+1)\alpha)} \int_a^x (x-t)^{(n+1)\alpha-1} \left(D_a^{(n+1)\alpha} f\right)(t) \, dt, \; \forall \, x \in [a,b], \text{ for } 0 < \alpha < 1. \tag{4.4.6}$$

Here we are going to operate more generally. Again we assume $0 < \alpha \le 1$, and $f : [a,b] \to X$, such that $f' \in C([a,b], X)$. We define the following X-valued left Caputo fractional derivatives:

$$\left(D_y^{\alpha} f\right)(x) = \frac{1}{\Gamma(1-\alpha)} \int_y^x (x-t)^{-\alpha} f'(t) \, dt, \tag{4.4.7}$$

for any $x \ge y$; $x, y \in [a,b]$, and

$$\left(D_x^{\alpha} f\right)(y) = \frac{1}{\Gamma(1-\alpha)} \int_x^y (y-t)^{-\alpha} f'(t) \, dt, \tag{4.4.8}$$

for any $y \ge x$; $x, y \in [a,b]$.

Notice $D_y^1 f = f'$, $D_x^1 f = f'$ by convention.

Clearly here $\left(D_y^{\alpha} f\right)$, $\left(D_x^{\alpha} f\right)$ are continuous functions over $[a,b]$, see [3]. We also make the convention that $\left(D_y^{\alpha} f\right)(x) = 0$, for $x < y$, and $\left(D_x^{\alpha} f\right)(y) = 0$, for $y < x$.

Here we assume that $D_y^{k\alpha} f$, $D_x^{k\alpha} f \in C^1([a,b], X)$, $k = 1, ..., n$; $D_y^{(n+1)\alpha} f$, $D_x^{(n+1)\alpha} f \in C([a,b], X)$, $n \in \mathbb{N}$; $\forall \, x, y \in [a,b]$.

Then by (4.4.6) we obtain

$$f(x) - f(y) = \sum_{i=2}^{n} \frac{(x-y)^{i\alpha}}{\Gamma(i\alpha+1)} \left(D_y^{i\alpha} f\right)(y) +$$

$$\frac{1}{\Gamma((n+1)\alpha)} \int_y^x (x-t)^{(n+1)\alpha-1} \left(D_y^{(n+1)\alpha} f\right)(t) \, dt, \tag{4.4.9}$$

$\forall \, x > y$; $x, y \in [a,b]$, for $0 < \alpha < 1$,

and also it holds

$$f(y) - f(x) = \sum_{i=2}^{n} \frac{(y-x)^{i\alpha}}{\Gamma(i\alpha+1)} \left(D_x^{i\alpha} f\right)(x) +$$

$$\frac{1}{\Gamma((n+1)\alpha)} \int_x^y (y-t)^{(n+1)\alpha-1} \left(D_x^{(n+1)\alpha} f\right)(t) \, dt, \tag{4.4.10}$$

$\forall \, y > x; \, x, y \in [a, b]$, for $0 < \alpha < 1$.

We define the following X-valued linear operator

$$(A(f))(x, y) =$$

$$\begin{cases} \sum_{i=2}^{n} \frac{(x-y)^{i\alpha-1}}{\Gamma(i\alpha+1)} \left(D_y^{i\alpha} f\right)(y) + \left(D_y^{(n+1)\alpha} f(x)\right) \frac{(x-y)^{(n+1)\alpha-1}}{\Gamma((n+1)\alpha+1)}, \, x > y, \\[2mm] \sum_{i=2}^{n} \frac{(y-x)^{i\alpha-1}}{\Gamma(i\alpha+1)} \left(D_x^{i\alpha} f\right)(x) + \left(D_x^{(n+1)\alpha} f(y)\right) \frac{(y-x)^{(n+1)\alpha-1}}{\Gamma((n+1)\alpha+1)}, \, y > x, \\[2mm] f'(x), \text{ when } x = y, \end{cases} \quad (4.4.11)$$

$\forall \, x, y \in [a, b], 0 < \alpha < 1$.

We may assume that

$$\|(A(f))(x, x) - (A(f))(y, y)\| = \|f'(x) - f'(y)\| \quad (4.4.12)$$

$$\leq \Phi \, |x - y|, \, \forall \, x, y \in [a, b], \text{ with } \Phi > 0,$$

see also ([12], p. 3).

We estimate and have:

i) case of $x > y$:

$$\|f(x) - f(y) - (A(f))(x, y)(x - y)\| =$$

$$\left\| \frac{1}{\Gamma((n+1)\alpha)} \int_y^x (x - t)^{(n+1)\alpha-1} \left(D_y^{(n+1)\alpha} f\right)(t) \, dt \right. \quad (4.4.13)$$

$$\left. - \left(D_y^{(n+1)\alpha} f(x)\right) \frac{(x - y)^{(n+1)\alpha}}{\Gamma((n+1)\alpha+1)} \right\|$$

(by [1], p. 426. Theorem 11.43)

$$= \frac{1}{\Gamma((n+1)\alpha)} \left\| \int_y^x (x - t)^{(n+1)\alpha-1} \left(\left(D_y^{(n+1)\alpha} f\right)(t) - \left(D_y^{(n+1)\alpha} f\right)(x)\right) dt \right\|$$

(by [8])

$$\leq \frac{1}{\Gamma((n+1)\alpha)} \int_y^x (x - t)^{(n+1)\alpha-1} \left\| D_y^{(n+1)\alpha} f(t) - \left(D_y^{(n+1)\alpha} f\right)(x) \right\| dt$$

(we assume here that

$$\left\| D_y^{(n+1)\alpha} f(t) - D_y^{(n+1)\alpha} f(x) \right\| \leq \lambda_1 \, |t - x|, \quad (4.4.14)$$

$\forall\, t, x, y \in [a, b] : x \geq t \geq y$, where $\lambda_1 > 0$)

$$\leq \frac{\lambda_1}{\Gamma((n+1)\alpha)} \int_y^x (x-t)^{(n+1)\alpha-1} (x-t)\, dt =$$

$$\frac{\lambda_1}{\Gamma((n+1)\alpha)} \int_y^x (x-t)^{(n+1)\alpha}\, dt = \frac{\lambda_1}{\Gamma((n+1)\alpha)} \frac{(x-y)^{(n+1)\alpha+1}}{((n+1)\alpha+1)}. \qquad (4.4.15)$$

We have proved that

$$\| f(x) - f(y) - (A(f))(x, y)(x-y) \| \leq \frac{\lambda_1}{\Gamma((n+1)\alpha)} \frac{(x-y)^{(n+1)\alpha+1}}{((n+1)\alpha+1)}, \tag{4.4.16}$$

for any $x, y \in [a, b] : x > y, 0 < \alpha < 1$.

(ii) case of $x < y$:

$$\| f(x) - f(y) - (A(f))(x, y)(x-y) \| =$$

$$\| f(y) - f(x) - (A(f))(x, y)(y-x) \| =$$

$$\left\| \frac{1}{\Gamma((n+1)\alpha)} \int_x^y (y-t)^{(n+1)\alpha-1} \left(D_x^{(n+1)\alpha} f\right)(t)\, dt \right. \tag{4.4.17}$$

$$\left. - \left(D_x^{(n+1)\alpha} f(y)\right) \frac{(y-x)^{(n+1)\alpha}}{\Gamma((n+1)\alpha+1)} \right\| =$$

$$\frac{1}{\Gamma((n+1)\alpha)} \left\| \int_x^y (y-t)^{(n+1)\alpha-1} \left(\left(D_x^{(n+1)\alpha} f\right)(t) - \left(D_x^{(n+1)\alpha} f\right)(y) \right) dt \right\| \leq$$

$$\frac{1}{\Gamma((n+1)\alpha)} \int_x^y (y-t)^{(n+1)\alpha-1} \left\| \left(D_x^{(n+1)\alpha} f\right)(t) - \left(D_x^{(n+1)\alpha} f\right)(y) \right\| dt$$

(we assume that

$$\left\| \left(D_x^{(n+1)\alpha} f\right)(t) - \left(D_x^{(n+1)\alpha} f\right)(y) \right\| \leq \lambda_2 |t - y|, \tag{4.4.18}$$

$\forall\, t, y, x \in [a, b] : y \geq t \geq x$, where $\lambda_2 > 0$)

$$\leq \frac{\lambda_2}{\Gamma((n+1)\alpha)} \int_x^y (y-t)^{(n+1)\alpha-1} (y-t)\, dt =$$

$$\frac{\lambda_2}{\Gamma((n+1)\alpha)} \int_x^y (y-t)^{(n+1)\alpha}\, dt = \frac{\lambda_2}{\Gamma((n+1)\alpha)} \frac{(y-x)^{(n+1)\alpha+1}}{((n+1)\alpha+1)}. \qquad (4.4.19)$$

We have proved that

$$\|f(x) - f(y) - A(f)(x,y)(x-y)\| \leq \frac{\lambda_2}{\Gamma((n+1)\alpha)} \frac{(y-x)^{(n+1)\alpha+1}}{((n+1)\alpha+1)},$$
(4.4.20)

$\forall x, y \in [a,b] : y > x, 0 < \alpha < 1$.

Conclusion Let $\lambda := \max(\lambda_1, \lambda_2)$. It holds

$$\|f(x) - f(y) - (A(f))(x,y)(x-y)\| \leq \frac{\lambda}{\Gamma((n+1)\alpha)} \frac{|x-y|^{(n+1)\alpha+1}}{((n+1)\alpha+1)},$$
(4.4.21)

$\forall x, y \in [a,b]$, where $0 < \alpha < 1, n \in \mathbb{N}$.

One may assume that $\frac{\lambda}{\Gamma((n+1)\alpha)} < 1$.

(Above notice that (4.4.21) is trivial when $x = y$.)

Now based on (4.4.12) and (4.4.21), we can apply our numerical methods presented in this chapter, to solve $f(x) = 0$.

To have $(n+1)\alpha + 1 \geq 2$, we need to take $1 > \alpha \geq \frac{1}{n+1}$, where $n \in \mathbb{N}$.

References

1. C.D. Aliprantis, K.C. Border, *Infinite Dimensional Analysis* (Springer, New York, 2006)
2. S. Amat, S. Busquier, S. Plaza, Chaotic dynamics of a third-order Newton-type method. J. Math. Anal. Applic. **366**(1), 164–174 (2010)
3. G.A. Anastassiou, A strong Fractional Calculus theory for Banach space valued functions, nonlinear functional analysis and applications (Korea) (2017). Accepted for publication
4. G.A. Anastassiou, *Principles of general fractional analysis for Banach space valued functions* (2017). submitted for publication
5. G.A. Anastassiou, I.K. Argyros, *Iterative convergence with Banach space valued functions in abstract fractional calculus* (2017). submitted
6. I.K. Argyros, A unifying local-semilocal convergence analysis and applications for two-point Newton-like methods in Banach space. J. Math. Anal. Appl. **298**, 374–397 (2004)
7. I.K. Argyros, A. Magréñan, *Iterative methods and their dynamics with applications* (CRC Press, New York, 2017)
8. Bochner integral. *Encyclopedia of Mathematics*, http://www.encyclopediaofmath.org/index.php?title=Bochner_integral&oldid=38659
9. M. Edelstein, On fixed and periodic points under contractive mappings. J. Lond. Math. Soc. **37**, 74–79 (1962)
10. J.A. Ezquerro, J.M. Gutierrez, M.A. Hernandez, N. Romero, M.J. Rubio, The Newton method: from Newton to Kantorovich (Spanish). Gac. R. Soc. Mat. Esp. **13**, 53–76 (2010)
11. L.V. Kantorovich, G.P. Akilov, *Functional Analysis in Normed Spaces* (Pergamon Press, New York, 1982)
12. G.E. Ladas, V. Lakshmikantham, *Differential equations in abstract spaces* (Academic Press, New York, London, 1972)
13. S. Maruster, Local convergence of Ezquerro-Hernandez method. Ann. West. Univ. Timisoara Math. Comput. Sci. **54**(1), 1–9 (2016)
14. J. Mikusinski, *The Bochner integral* (Academic Press, New York, 1978)

15. F.A. Potra, V. Pták, *Nondiscrete induction and iterative processes* (Pitman Publication, London, 1984)
16. P.D. Proinov, New general convergence theory for iterative processes and its applications to Newton-Kantorovich type theorems. J. Complexity **26**, 3–42 (2010)
17. G.E. Shilov, *Elementary Functional Analysis* (Dover Publications Inc, New York, 1996)

Chapter 5
Semi-local Convergence in Right Abstract Fractional Calculus

We provide a semi-local convergence analysis for a class of iterative methods under generalized conditions in order to solve equations in a Banach space setting. Some applications are suggested including Banach space valued functions of right fractional calculus, where all integrals are of Bochner-type. It follows [5].

5.1 Introduction

Sections 5.1–5.3 are prerequisites for Sect. 5.4.

Let B_1, B_2 stand for Banach spaces and let Ω stand for an open subset of B_1. Let also $U(z, \xi) := \{u \in B_1 : \|u - z\| < \xi\}$ and let $\overline{U}(z, \xi)$ stand for the closure of $U(z, \xi)$.

Many problems in Computational Sciences, Engineering, Mathematical Chemistry, Mathematical Physics, Mathematical Economics and other disciplines can be brought in a form like

$$F(x) = 0 \tag{5.1.1}$$

using Mathematical Modeling [1–18], where $F : \Omega \rightarrow B_2$ is a continuous operator. The solution x^* of Eq. (5.1.1) is needed in closed form. This is possible only in special cases, which explains why most solution methods for such equations are usually iterative. There is a plethora of iterative methods for solving Eq. (5.1.1). We can divide these methods in two categories.

Explicit Methods [7, 8, 12, 16, 17]: Newton's method

$$x_{n+1} = x_n - F'(x_n)^{-1} F(x_n). \tag{5.1.2}$$

Secant method:

$$x_{n+1} = x_n - \left[x_{n-1}, x_n; F\right]^{-1} F(x_n), \tag{5.1.3}$$

© Springer International Publishing AG 2018
G. A. Anastassiou and I. K. Argyros, *Functional Numerical Methods: Applications to Abstract Fractional Calculus*, Studies in Systems, Decision and Control 130, https://doi.org/10.1007/978-3-319-69526-6_5

where $[\cdot, \cdot; F]$ denotes a divided difference of order one on $\Omega \times \Omega$ [8, 16, 17].

Newton-like method:

$$x_{n+1} = x_n - E_n^{-1} F(x_n), \qquad (5.1.4)$$

where $E_n = E(F)(x_n)$ and $E : \Omega \to \mathcal{L}(B_1, B_2)$ the space of bounded linear operators from B_1 into B_2. Other explicit methods can be found in [8, 12, 16, 17] and the references there in.

Implicit Methods [7, 10, 12, 17]:

$$F(x_n) + A_n(x_{n+1} - x_n) = 0 \qquad (5.1.5)$$

$$x_{n+1} = x_n - A_n^{-1} F(x_n), \qquad (5.1.6)$$

where $A_n = A(x_{n+1}, x_n) = A(F)(x_{n+1}, x_n)$ and $A : \Omega \times \Omega \to \mathcal{L}(B_1, B_2)$. We let $A(F)(x, x) = A(x, x) = A(x)$ for each $x \in \Omega$.

There is a plethora on local as well as semi-local convergence results for explicit methods [1–9, 11–17]. However, the research on the convergence of implicit methods has received little attention. Authors, usually consider the fixed point problem

$$P_z(x) = x, \qquad (5.1.7)$$

where

$$P_z(x) = x + F(z) + A(x, z)(x - z) \qquad (5.1.8)$$

or

$$P_z(x) = z - A(x, z)^{-1} F(z) \qquad (5.1.9)$$

for methods (5.1.5) and (5.1.6), respectivelly, where $z \in \Omega$ is given. If P is a contraction operator mapping a closed set into itself, then according to the contraction mapping principle [12, 16, 17], P_z has a fixed point x_z^* which can be found using the method of succesive substitutions or Picard's method [17] defined for each fixed n by

$$y_{k+1,n} = P_{x_n}(y_{k,n}), \quad y_{0,n} = x_n, \ x_{n+1} = \lim_{k \to +\infty} y_{k,n}. \qquad (5.1.10)$$

Let us also consider the analogous explicit methods

$$F(x_n) + A(x_n, x_n)(x_{n+1} - x_n) = 0 \qquad (5.1.11)$$

$$x_{n+1} = x_n - A(x_n, x_n)^{-1} F(x_n) \qquad (5.1.12)$$

$$F(x_n) + A(x_n, x_{n-1})(x_{n+1} - x_n) = 0 \qquad (5.1.13)$$

and

$$x_{n+1} = x_n - A(x_n, x_{n-1})^{-1} F(x_n). \qquad (5.1.14)$$

In this chapter in Sect. 5.2, we present the semi-local convergence of method (5.1.5) and method (5.1.6). Section 5.3 contains the semi-local convergence of method (5.1.11), method (5.1.12), method (5.1.13) and method (5.1.14). Some applications to Abstract Fractional Calculus are suggested in Sect. 5.4 on a certain Banach space valued functions, where all the integrals are of Bochner-type [9, 15].

5.2 Semi-local Convergence for Implicit Methods

We present the semi-local convergence analysis of method (5.1.6) using conditions (S):

(s_1) $F : \Omega \subset B_1 \rightarrow B_2$ is continuous and $A(x, y) \in \mathcal{L}(B_1, B_2)$ for each $(x, y) \in \Omega \times \Omega$.

(s_2) There exist $\beta > 0$ and $\Omega_0 \subset B_1$ such that $A(x, y)^{-1} \in \mathcal{L}(B_2, B_1)$ for each $(x, y) \in \Omega_0 \times \Omega_0$ and

$$\left\| A(x, y)^{-1} \right\| \leq \beta^{-1}.$$

Set $\Omega_1 = \Omega \cap \Omega_0$.

(s_3) There exists a continuous and nondecreasing function $\psi : [0, +\infty)^3 \rightarrow [0, +\infty)$ such that for each $x, y \in \Omega_1$

$$\| F(x) - F(y) - A(x, y)(x - y) \| \leq$$

$$\beta \psi(\|x - y\|, \|x - x_0\|, \|y - x_0\|) \|x - y\|.$$

(s_4) For each $x \in \Omega_0$ there exists $y \in \Omega_0$ such that

$$y = x - A(y, x)^{-1} F(x).$$

(s_5) For $x_0 \in \Omega_0$ and $x_1 \in \Omega_0$ satisfying (s_4) there exists $\eta \geq 0$ such that

$$\left\| A(x_1, x_0)^{-1} F(x_0) \right\| \leq \eta.$$

(s_6) Define $q(t) := \psi(\eta, t, t)$ for each $t \in [0, +\infty)$. Equation

$$t(1 - q(t)) - \eta = 0$$

has positive solutions. Denote by s the smallest such solution.

(s_7) $\overline{U}(x_0, s) \subset \Omega$, where

$$s = \frac{\eta}{1 - q_0} \quad \text{and} \quad q_0 = \psi(\eta, s, s).$$

Next, we present the semi-local convergence analysis for method (5.1.6) using the conditions (S) and the preceding notation.

Theorem 5.1 *Assume that the conditions (S) hold. Then, sequence $\{x_n\}$ generated by method (5.1.6) starting at $x_0 \in \Omega$ is well defined in $U(x_0, s)$, remains in $U(x_0, s)$ for each $n = 0, 1, 2, ...$ and converges to a solution $x^* \in \overline{U}(x_0, s)$ of equation $F(x) = 0$. Moreover, suppose that there exists a continuous and nondecreasing function $\psi_1 : [0, +\infty)^4 \rightarrow [0, +\infty)$ such that for each $x, y, z \in \Omega_1$*

$$\|F(x) - F(y) - A(z, y)(x - y)\| \leq$$

$$\beta\psi_1(\|x - y\|, \|x - x_0\|, \|y - x_0\|, \|z - x_0\|)\|x - y\|$$

and $q_1 = \psi_1(\eta, s, s, s) < 1$.
    Then, x^* is the unique solution of equation $F(x) = 0$ in $\overline{U}(x_0, s)$.

Proof By the definition of s and (s_5), we have $x_1 \in U(x_0, s)$. The proof is based on mathematical induction on k. Suppose that $\|x_k - x_{k-1}\| \leq q_0^{k-1}\eta$ and $\|x_k - x_0\| \leq s$.
    We get by (5.1.6), $(s_2) - (s_5)$ in turn that

$$\|x_{k+1} - x_k\| = \left\|A_k^{-1}F(x_k)\right\| = \left\|A_k^{-1}(F(x_k) - F(x_{k-1}) - A_{k-1}(x_k - x_{k-1}))\right\|$$

$$\leq \left\|A_k^{-1}\right\| \|F(x_k) - F(x_{k-1}) - A_{k-1}(x_k - x_{k-1})\| \leq$$

$$\beta^{-1}\beta\psi(\|x_k - x_{k-1}\|, \|x_{k-1} - x_0\|, \|y_k - x_0\|)\|x_k - x_{k-1}\| \leq$$

$$\psi(\eta, s, s)\|x_k - x_{k-1}\| = q_0\|x_k - x_{k-1}\| \leq q_0^k\|x_1 - x_0\| \leq q_0^k\eta \qquad (5.2.1)$$

and

$$\|x_{k+1} - x_0\| \leq \|x_{k+1} - x_k\| + \cdots + \|x_1 - x_0\|$$

$$\leq q_0^k\eta + \cdots + \eta = \frac{1 - q_0^{k+1}}{1 - q_0}\eta < \frac{\eta}{1 - q_0} = s.$$

The induction is completed. Moreover, we have by (5.2.1) that for $m = 0, 1, 2, ...$

$$\|x_{k+m} - x_k\| \leq \frac{1 - q_0^m}{1 - q_0}q_0^k\eta.$$

It follows from the preceding inequation that sequence $\{x_k\}$ is complete in a Banach space B_1 and as such it converges to some $x^* \in \overline{U}(x_0, s)$ (since $\overline{U}(x_0, s)$ is a closed ball). By letting $k \rightarrow +\infty$ in (5.2.1) we get $F(x^*) = 0$. To show the uniqueness part, let $x^{**} \in U(x_0, s)$ be a solution of equation $F(x) = 0$. By using (5.1.6) and the hypothesis on ψ_1, we obtain in turn that

$$\left\|x^{**} - x_{k+1}\right\| = \left\|x^{**} - x_k + A_k^{-1}F(x_k) - A_k^{-1}F(x^{**})\right\| \leq$$

$$\left\| A_k^{-1} \right\| \left\| F\left(x^{**}\right) - F\left(x_k\right) - A_k\left(x^{**} - x_k\right) \right\| \leq$$

$$\beta^{-1}\beta\psi_1 \left(\left\| x^{**} - x_k \right\|, \left\| x_{k-1} - x_0 \right\|, \left\| x_k - x_0 \right\|, \left\| x^{**} - x_0 \right\| \right) \left\| x^{**} - x_k \right\| \leq$$

$$q_1 \left\| x^{**} - x_k \right\| \leq q_1^{k+1} \left\| x^{**} - x_0 \right\|,$$

so $\lim_{k \to +\infty} x_k = x^{**}$. We have shown that $\lim_{k \to +\infty} x_k = x^*$, so $x^* = x^{**}$. ∎

Remark 5.2 (1) The equation in (s_6) is used to determine the smallness of η. It can be replaced by a stronger condition as follows. Choose $\mu \in (0, 1)$. Denote by s_0 the smallest positive solution of equation $q(t) = \mu$. Notice that if function q is strictly increasing, we can set $s_0 = q^{-1}(\mu)$. Then, we can suppose instead of (s_6) :

(s_6') $\eta \leq (1 - \mu) s_0$

which is a stronger condition than (s_6).

However, we wanted to leave the equation in (s_6) as uncluttered and as weak as possible.

(2) Condition (s_2) can become part of condition (s_3) by considering

(s_3)' There exists a continuous and nondecreasing function $\varphi : [0, +\infty)^3 \to [0, +\infty)$ such that for each $x, y \in \Omega_1$

$$\left\| A(x, y)^{-1} \left[F(x) - F(y) - A(x, y)(x, y) \right] \right\| \leq$$

$$\varphi\left(\left\| x - y \right\|, \left\| x - x_0 \right\|, \left\| y - x_0 \right\| \right) \left\| x - y \right\|.$$

Notice that

$$\varphi(u_1, u_2, u_3) \leq \psi(u_1, u_2, u_3)$$

for each $u_1 \geq 0$, $u_2 \geq 0$ and $u_3 \geq 0$. Similarly, a function φ_1 can replace ψ_1 for the uniqueness of the solution part. These replacements are of Mysovskii-type [7, 12, 16] and influence the weaking of the convergence criterion in (s_6), error bounds and the precision of s.

(3) Suppose that there exist $\beta > 0$, $\beta_1 > 0$ and $L \in \mathcal{L}(B_1, B_2)$ with $L^{-1} \in \mathcal{L}(B_2, B_1)$ such that

$$\left\| L^{-1} \right\| \leq \beta^{-1}$$

$$\left\| A(x, y) - L \right\| \leq \beta_1$$

and

$$\beta_2 := \beta^{-1}\beta_1 < 1.$$

Then, it follows from the Banach lemma on invertible operators [12], and

$$\left\| L^{-1} \right\| \left\| A(x, y) - L \right\| \leq \beta^{-1}\beta_1 = \beta_2 < 1$$

that $A(x, y)^{-1} \in \mathcal{L}(B_2, B_1)$. Let $\beta = \frac{\beta^{-1}}{1 - \beta_2}$. Then, under these replacements, condition (s_2) is implied, therefore it can be dropped from the conditions (S).

(4) Clearly method (5.1.5) converges under the conditions (S), since (5.1.6) implies (5.1.5).

(5) We wanted to leave condition (s_4) as uncluttered as possible, since in practice equations (5.1.6) (or (5.1.5)) may be solvable in a way avoiding the already mentioned conditions of the contraction mapping principle. However, in what follows we examine the solvability of method (5.1.5) under a stronger version of the contraction mapping principle using the conditions (V) :

$(v_1) = (s_1)$.

(v_2) There exist functions $w_1 : [0, +\infty)^4 \to [0, +\infty)$, $w_2 : [0, +\infty)^4 \to [0, +\infty)$ continuous and nondecreasing such that for each $x, y, z \in \Omega$

$$\|I + A(x, z) - A(y, z)\| \leq w_1(\|x - y\|, \|x - x_0\|, \|y - x_0\|, \|z - x_0\|)$$

$$\|A(x, z) - A(y, z)\| \leq w_2(\|x - y\|, \|x - x_0\|, \|y - x_0\|, \|z - x_0\|) \|x - y\|$$

and

$$w_1(0, 0, 0, 0) = w_2(0, 0, 0, 0) = 0.$$

Set

$$h(t, t, t, t) = \begin{cases} w_1(2t, t, t, t) + w_2(2t, t, t, t)(t + \|x_0\|), \ z \neq x_0 \\ w_1(2t, t, t, 0) + w_2(2t, t, t, 0) \|x_0\|, \ z = x_0. \end{cases}$$

(v_3) There exists $\tau > 0$ satisfying

$$h(t, t, t, t) < 1$$

and

$$h(t, t, 0, t)t + \|F(x_0)\| \leq t$$

(v_4) $\overline{U}(x_0, \tau) \subseteq D$.

Theorem 5.3 *Suppose that the conditions (V) are satisfied. Then, equation (5.1.5) is uniquely solvable for each $n = 0, 1, 2, \ldots$. Moreover, if $A_n^{-1} \in \mathcal{L}(B_2, B_1)$, the Eq. (5.1.6) is also uniquely solvable for each $n = 0, 1, 2, \ldots$*

Proof The result is an application of the contraction mapping principle. Let $x, y, z \in U(x_0, \tau)$. By the definition of operator P_z, (v_2) and (v_3), we get in turn that

$$\|P_z(x) - P_z(y)\| = \|(I + A(x, z) - A(y, z))(x - y) - (A(x, z) - A(y, z))z\|$$

$$\leq \|I + A(x, z) - A(y, z)\| \|x - y\| + \|A(x, z) - A(y, z)\| \|z\|$$

$$\leq [w_1(\|x - y\|, \|x - x_0\|, \|y - x_0\|, \|z - x_0\|) +$$

$$w_2 \left(\|x - y\|, \|x - x_0\|, \|y - x_0\|, \|z - x_0\| \right) \left(\|z - x_0\| + \|x_0\| \right) \|x - y\|$$

$$\leq h \left(\tau, \tau, \tau, \tau \right) \|x - y\|$$

and

$$\|P_z(x) - x_0\| \leq \|P_z(x) - P_z(x_0)\| + \|P_z(x_0) - x_0\|$$

$$\leq h \left(\|x - x_0\|, \|x - x_0\|, 0, \|z - x_0\| \right) \|x - x_0\| + \|F(x_0)\|$$

$$\leq h \left(\tau, \tau, 0, \tau \right) \tau + \|F(x_0)\| \leq \tau. \qquad \blacksquare$$

Remark 5.4 Sections 5.2 and 5.3 have an interest independent of Sect. 5.4. It is worth noticing that the results especially of Theorem 5.1 can apply in Abstract Fractional Calculus as illustrated in Sect. 5.4. By specializing function ψ, we can apply the results of say Theorem 5.1 in the examples suggested in Sect. 5.4. In particular for (5.4.28), we choose $\psi(u_1, u_2, u_3) = \frac{\lambda u_1^{(n+1)\alpha}}{\beta \Gamma((n+1)\alpha)((n+1)\alpha+1)}$ for $u_1 \geq 0, u_2 \geq 0, u_3 \geq 0$ and λ, α are given in Sect. 5.4. Similar choices for the other examples of Sect. 5.4. It is also worth noticing that estimate (5.4.2) derived in Sect. 5.4 is of independent interest but not needed in Theorem 5.1.

5.3   Semi-local Convergence for Explicit Methods

A specialization of Theorem 5.1 can be utilized to study the semi-local convergence of the explicit methods given in the introduction of this study. In particular, for the study of method (5.1.12) (and consequently of method (5.1.11)), we use the conditions (S') :

(s'_1) $F : \Omega \subset B_1 \rightarrow B_2$ is continuous and $A(x, x) \in \mathcal{L}(B_1, B_2)$ for each $x \in \Omega$.

(s'_2) There exist $\beta > 0$ and $\Omega_0 \subset B_1$ such that $A(x, x)^{-1} \in \mathcal{L}(B_2, B_1)$ for each $x \in \Omega_0$ and

$$\left\| A(x, x)^{-1} \right\| \leq \beta^{-1}.$$

Set $\Omega_1 = \Omega \cap \Omega_0$.

(s'_3) There exist continuous and nondecreasing functions $\psi_0 : [0, +\infty)^3 \rightarrow [0, +\infty)$, $\psi_2 : [0, +\infty)^3 \rightarrow [0, +\infty)$ with $\psi_0(0, 0, 0) = \psi_2(0, 0, 0) = 0$ such that for each $x, y \in \Omega_1$

$$\|F(x) - F(y) - A(y, y)(x - y)\| \leq$$

$$\beta \psi_0 \left(\|x - y\|, \|x - x_0\|, \|y - x_0\| \right) \|x - y\|$$

and

$$\|A(x, y) - A(y, y)\| \leq \beta \psi_2 \left(\|x - y\|, \|x - x_0\|, \|y - x_0\| \right).$$

Set $\psi = \psi_0 + \psi_2$.

(s_4') There exist $x_0 \in \Omega_0$ and $\eta \geq 0$ such that $A(x_0, x_0)^{-1} \in \mathcal{L}(B_2, B_1)$ and

$$\left\| A(x_0, x_0)^{-1} F(x_0) \right\| \leq \eta.$$

$(s_5') = (s_6)$
$(s_6') = (s_7)$.

Next, we present the following semi-local convergence analysis of method (5.1.12) using the (S') conditions and the preceding notation.

Proposition 5.5 *Suppose that the conditions (S') are satisfied. Then, sequence $\{x_n\}$ generated by method (5.1.12) starting at $x_0 \in \Omega$ is well defined in $U(x_0, s)$, remains in $U(x_0, s)$ for each $n = 0, 1, 2, \dots$ and converges to a unique solution $x^* \in \overline{U}(x_0, s)$ of equation $F(x) = 0$.*

Proof We follow the proof of Theorem 5.1 but use instead the analogous estimate

$$\| F(x_k) \| = \| F(x_k) - F(x_{k-1}) - A(x_{k-1}, x_{k-1})(x_k - x_{k-1}) \| \leq$$

$$\| F(x_k) - F(x_{k-1}) - A(x_k, x_{k-1})(x_k - x_{k-1}) \| +$$

$$\| (A(x_k, x_{k-1}) - A(x_{k-1}, x_{k-1}))(x_k - x_{k-1}) \| \leq$$

$$\left[\psi_0 \left(\| x_k - x_{k-1} \|, \| x_{k-1} - x_0 \|, \| x_k - x_0 \| \right) + \right.$$

$$\left. \psi_2 \left(\| x_k - x_{k-1} \|, \| x_{k-1} - x_0 \|, \| x_k - x_0 \| \right) \right] \| x_k - x_{k-1} \| =$$

$$\psi \left(\| x_k - x_{k-1} \|, \| x_{k-1} - x_0 \|, \| x_k - x_0 \| \right) \| x_k - x_{k-1} \|.$$

The rest of the proof is identical to the one in Theorem 5.1 until the uniqueness part for which we have the corresponding estimate

$$\left\| x^{**} - x_{k+1} \right\| = \left\| x^{**} - x_k + A_k^{-1} F(x_k) - A_k^{-1} F(x^{**}) \right\| \leq$$

$$\left\| A_k^{-1} \right\| \left\| F(x^{**}) - F(x_k) - A_k(x^{**} - x_k) \right\| \leq$$

$$\beta^{-1} \bar{\beta} \psi_0 \left(\left\| x^{**} - x_k \right\|, \| x_{k-1} - x_0 \|, \| x_k - x_0 \| \right) \leq$$

$$q \left\| x^{**} - x_k \right\| \leq q^{k+1} \left\| x^{**} - x_0 \right\|. \qquad \blacksquare$$

Remark 5.6 Comments similar to the ones given in Sect. 5.2 can follows but for method (5.1.13) and method (5.1.14) instead of method (5.1.5) and method (5.1.6), respectively.

5.4 Applications to *X*-valued Right Fractional Calculus

Here we deal with Banach space $(X, \|\cdot\|)$ valued functions f of real domain $[a, b]$. All integrals here are of Bochner-type, see [15]. The derivatives of f are defined similarly to numerical ones, see [18], pp. 83–86 and p. 93.

Let $f : [a, b] \rightarrow X$ such that $f^{(m)} \in L_\infty ([a, b], X)$. The *X*-valued right Caputo fractional derivative of order $\alpha \notin \mathbb{N}$, $\alpha > 0$, $m = \lceil \alpha \rceil$ ($\lceil \cdot \rceil$ ceiling), is defined as follows (see [3]):

$$\left(D_{b-}^\alpha f \right) (x) := \frac{(-1)^m}{\Gamma (m - \alpha)} \int_x^b (z - x)^{m-\alpha-1} f^{(m)} (z) \, dz, \qquad (5.4.1)$$

$\forall\, x \in [a, b]$, with $D_{b-}^m f (x) := (-1)^m f^{(m)} (x)$, $D_{b-}^0 f := f$, where Γ is the gamma function.

We observe that

$$\left\| \left(D_{b-}^\alpha f \right) (x) \right\| \leq \frac{1}{\Gamma (m - \alpha)} \int_x^b (z - x)^{m-\alpha-1} \left\| f^{(m)} (z) \right\| dz$$

$$\leq \frac{\left\| f^{(m)} \right\|_\infty}{\Gamma (m - \alpha)} \left(\int_x^b (z - x)^{m-\alpha-1} dz \right) = \frac{\left\| f^{(m)} \right\|_\infty}{\Gamma (m - \alpha)} \frac{(b - x)^{m-\alpha}}{m - \alpha} \qquad (5.4.2)$$

$$= \frac{\left\| f^{(m)} \right\|_\infty (b - x)^{m-\alpha}}{\Gamma (m - \alpha + 1)}.$$

We have proved that

$$\left\| \left(D_{b-}^\alpha f \right) (x) \right\| \leq \frac{\left\| f^{(m)} \right\|_\infty (b - x)^{m-\alpha}}{\Gamma (m - \alpha + 1)} \leq \frac{\left\| f^{(m)} \right\|_\infty (b - a)^{m-\alpha}}{\Gamma (m - \alpha + 1)}. \qquad (5.4.3)$$

Clearly here $\left(D_{b-}^\alpha f \right) (b) = 0$, $0 < \alpha \notin \mathbb{N}$.

Let $n \in \mathbb{N}$. We denote

$$D_{b-}^{n\alpha} := D_{b-}^\alpha D_{b-}^\alpha ... D_{b-}^\alpha \quad (n \text{ - times}). \qquad (5.4.4)$$

The *X*-valued right Riemann-Liouville fractional integral of order α, is defined as follows:

$$\left(I_{b-}^\alpha f \right) (x) := \frac{1}{\Gamma (\alpha)} \int_x^b (z - x)^{\alpha-1} f (z) \, dz, \qquad (5.4.5)$$

$\forall\, x \in [a, b]$, $I_{b-}^0 := I$ (the identity operator).

We denote also

$$I_{b-}^{n\alpha} := I_{b-}^\alpha I_{b-}^\alpha ... I_{b-}^\alpha \quad (n \text{ - times}). \qquad (5.4.6)$$

From now on we assume $0 < \alpha \leq 1$, that is $m = 1$.

In [4], we proved the following X-valued right generalized fractional Taylor's formula:

Theorem 5.7 *Suppose that* $f \in C^1([a, b], X)$ *and* $D_{b-}^{k\alpha} f \in C^1([a, b], X)$, *for* $k = 1, ..., n \in \mathbb{N}$; $D_{b-}^{(n+1)\alpha} f \in C([a, b], X)$, *where* $0 < \alpha \leq 1$. *Then*

$$f(x) = \sum_{i=0}^{n} \frac{(b-x)^{i\alpha}}{\Gamma(i\alpha+1)} \left(D_{b-}^{i\alpha} f\right)(b) + \tag{5.4.7}$$

$$\frac{1}{\Gamma((n+1)\alpha)} \int_x^b (z-x)^{(n+1)\alpha-1} \left(D_{b-}^{(n+1)\alpha} f\right)(z) \, dz, \quad \forall \, x \in [a, b].$$

We make

Remark 5.8 In particular, when $f' \in L_\infty([a, b], X)$, $0 < \alpha < 1$, we have that $D_{b-}^{\alpha} f(b) = 0$, also $\left(D_{b-}^1 f\right)(x) = -f'(x)$, and

$$\left(D_{b-}^{\alpha} f\right)(x) = \frac{-1}{\Gamma(1-\alpha)} \int_x^b (z-x)^{-\alpha} f'(z) \, dz, \quad \forall \, x \in [a, b]. \tag{5.4.8}$$

Thus, from (5.4.7) we derive

$$f(x) - f(b) = \sum_{i=2}^{n} \frac{(b-x)^{i\alpha}}{\Gamma(i\alpha+1)} \left(D_{b-}^{i\alpha} f\right)(b) + \tag{5.4.9}$$

$$\frac{1}{\Gamma((n+1)\alpha)} \int_x^b (z-x)^{(n+1)\alpha-1} \left(D_{b-}^{(n+1)\alpha} f\right)(z) \, dz, \quad \forall \, x \in [a, b]; 0 < \alpha < 1.$$

Here we are going to operate more generally. Again we assume $0 < \alpha \leq 1$, and $f : [a, b] \to X$, such that $f' \in C([a, b], X)$. We define the following X-valued right Caputo fractional derivatives:

$$\left(D_{y-}^{\alpha} f\right)(x) := \frac{-1}{\Gamma(1-\alpha)} \int_x^y (t-x)^{-\alpha} f'(t) \, dt, \tag{5.4.10}$$

for any $x \leq y$; $x, y \in [a, b]$, and

$$\left(D_{x-}^{\alpha} f\right)(y) = \frac{-1}{\Gamma(1-\alpha)} \int_y^x (t-y)^{-\alpha} f'(t) \, dt, \tag{5.4.11}$$

for any $y \leq x$; $x, y \in [a, b]$.

Notice $D_{y-}^1 f = -f'$, $D_{x-}^1 f = -f'$, by convention.

Clearly here $D_{y-}^\alpha f$, $D_{x-}^\alpha f$ are continuous functions over $[a, b]$, see [3]. We also make the convention that $\left(D_{y-}^\alpha f\right)(x) = 0$, for $x > y$, and $\left(D_{x-}^\alpha f\right)(y) = 0$, for $y > x$.

Here we assume that

$$D_{y-}^{k\alpha} f, D_{x-}^{k\alpha} f \in C^1\left([a, b], X\right), \tag{5.4.12}$$

$k = 0, 1, ..., n, n \in \mathbb{N}$ and $D_{y-}^{(n+1)\alpha} f, D_{x-}^{(n+1)\alpha} f \in C\left([a, b], X\right)$; $\forall x, y \in [a, b]$; and $0 < \alpha < 1$.

By (5.4.9) we derive the X-valued formulae:

$$f(x) - f(y) = \sum_{i=2}^{n} \frac{(y-x)^{i\alpha}}{\Gamma(i\alpha+1)} \left(D_{y-}^{i\alpha} f\right)(y) +$$

$$\frac{1}{\Gamma((n+1)\alpha)} \int_x^y (z-x)^{(n+1)\alpha-1} \left(D_{y-}^{(n+1)\alpha} f\right)(z)\, dz, \tag{5.4.13}$$

$\forall x < y$; $x, y \in [a, b]$; $0 < \alpha < 1$, and also it holds

$$f(y) - f(x) = \sum_{i=2}^{n} \frac{(x-y)^{i\alpha}}{\Gamma(i\alpha+1)} \left(D_{x-}^{i\alpha} f\right)(x) +$$

$$\frac{1}{\Gamma((n+1)\alpha)} \int_y^x (z-y)^{(n+1)\alpha-1} \left(D_{x-}^{(n+1)\alpha} f\right)(z)\, dz, \tag{5.4.14}$$

$\forall y < x$; $x, y \in [a, b]$; $0 < \alpha < 1$.

We define the following X-valued linear operator

$$(A(f))(x, y) :=$$

$$\begin{cases} \sum_{i=2}^{n} \frac{(y-x)^{i\alpha-1}}{\Gamma(i\alpha+1)} \left(D_{y-}^{i\alpha} f\right)(y) - \left(D_{y-}^{(n+1)\alpha} f(x)\right) \frac{(y-x)^{(n+1)\alpha-1}}{\Gamma((n+1)\alpha+1)}, & x < y, \\ \sum_{i=2}^{n} \frac{(x-y)^{i\alpha-1}}{\Gamma(i\alpha+1)} \left(D_{x-}^{i\alpha} f\right)(x) - \left(D_{x-}^{(n+1)\alpha} f(y)\right) \frac{(x-y)^{(n+1)\alpha-1}}{\Gamma((n+1)\alpha+1)}, & x > y, \\ f'(x), & \text{when } x = y, \end{cases} \tag{5.4.15}$$

$\forall x, y \in [a, b]$; $0 < \alpha < 1$.

We may assume that

$$\|(A(f))(x, x) - (A(f))(y, y)\| = \|f'(x) - f'(y)\| \leq \Phi |x - y|, \ \forall x, y \in [a, b], \tag{5.4.16}$$

(see [13], p. 3), with $\Phi > 0$.

We estimate and have:
(i) case $x < y$:

$$\|f(x) - f(y) - (A(f))(x, y)(x - y)\| =$$

$$\|f(y) - f(x) - (A(f))(x, y)(y - x)\| = \qquad (5.4.17)$$

$$\left\| \frac{1}{\Gamma((n+1)\alpha)} \int_x^y (z - x)^{(n+1)\alpha-1} \left(D_{y-}^{(n+1)\alpha} f\right)(z) \, dz - \left(D_{y-}^{(n+1)\alpha} f(x)\right) \frac{(y - x)^{(n+1)\alpha}}{\Gamma((n+1)\alpha + 1)} \right\|$$

(by [1], p. 426, Theorem 11.43)

$$= \frac{1}{\Gamma((n+1)\alpha)} \left\| \int_x^y (z - x)^{(n+1)\alpha-1} \left(D_{y-}^{(n+1)\alpha} f(z) - D_{y-}^{(n+1)\alpha} f(x)\right) dz \right\| \qquad (5.4.18)$$

(by [9])

$$\leq \frac{1}{\Gamma((n+1)\alpha)} \left(\int_x^y (z - x)^{(n+1)\alpha-1} \left\| D_{y-}^{(n+1)\alpha} f(z) - D_{y-}^{(n+1)\alpha} f(x) \right\| dz \right)$$

(we assume here that

$$\left\| D_{y-}^{(n+1)\alpha} f(z) - D_{y-}^{(n+1)\alpha} f(x) \right\| \leq \lambda_1 |z - x|, \qquad (5.4.19)$$

$\forall z, x, y \in [a, b] : y \geq z \geq x; \lambda_1 > 0$)

$$\leq \frac{\lambda_1}{\Gamma((n+1)\alpha)} \int_x^y (z - x)^{(n+1)\alpha-1} (z - x) \, dz =$$

$$\frac{\lambda_1}{\Gamma((n+1)\alpha)} \int_x^y (z - x)^{(n+1)\alpha} \, dz = \frac{\lambda_1}{\Gamma((n+1)\alpha)} \frac{(y - x)^{(n+1)\alpha+1}}{((n+1)\alpha + 1)}. \qquad (5.4.20)$$

We have proved that

$$\|f(x) - f(y) - (A(f))(x, y)(x - y)\| \leq \frac{\lambda_1 (y - x)^{(n+1)\alpha+1}}{\Gamma((n+1)\alpha)((n+1)\alpha + 1)}, \qquad (5.4.21)$$

for any $x, y \in [a, b] : x < y; 0 < \alpha < 1$.
(ii) Case of $x > y$: We have

$$\|f(y) - f(x) - (A(f))(x, y)(y - x)\| = \qquad (5.4.22)$$

$$\left\| \frac{1}{\Gamma\left((n+1)\,\alpha\right)} \int_y^x (z-y)^{(n+1)\alpha-1} \left(D_{x-}^{(n+1)\alpha} f\right)(z)\, dz - \right.$$

$$\left. \left(D_{x-}^{(n+1)\alpha} f\,(y)\right) \frac{(x-y)^{(n+1)\alpha}}{\Gamma\left((n+1)\,\alpha+1\right)} \right\| =$$

$$\frac{1}{\Gamma\left((n+1)\,\alpha\right)} \left\| \int_y^x (z-y)^{(n+1)\alpha-1} \left(\left(D_{x-}^{(n+1)\alpha} f\right)(z) - \left(D_{x-}^{(n+1)\alpha} f\right)(y)\right) dz \right\| \le$$

(5.4.23)

$$\frac{1}{\Gamma\left((n+1)\,\alpha\right)} \int_y^x (z-y)^{(n+1)\alpha-1} \left\| \left(D_{x-}^{(n+1)\alpha} f\right)(z) - \left(D_{x-}^{(n+1)\alpha} f\right)(y) \right\| dz$$

(we assume that

$$\left\| \left(D_{x-}^{(n+1)\alpha} f\right)(z) - \left(D_{x-}^{(n+1)\alpha} f\right)(y) \right\| \le \lambda_2 \,|z-y|,$$ (5.4.24)

$\forall\, z, y, x \in [a,b] : x \ge z \ge y;\ \lambda_2 > 0$)

$$\le \frac{\lambda_2}{\Gamma\left((n+1)\,\alpha\right)} \int_y^x (z-y)^{(n+1)\alpha-1}\,(z-y)\, dz =$$

$$\frac{\lambda_2}{\Gamma\left((n+1)\,\alpha\right)} \int_y^x (z-y)^{(n+1)\alpha}\, dz = \frac{\lambda_2}{\Gamma\left((n+1)\,\alpha\right)} \frac{(x-y)^{(n+1)\alpha+1}}{\left((n+1)\,\alpha+1\right)}.$$ (5.4.25)

We have proved that

$$\| f(x) - f(y) - (A(f))(x,y)(x-y) \| \le \frac{\lambda_2}{\Gamma\left((n+1)\,\alpha\right)} \frac{(x-y)^{(n+1)\alpha+1}}{\left((n+1)\,\alpha+1\right)},$$

(5.4.26)

for any $x, y \in [a,b] : x > y;\ 0 < \alpha < 1$.

Conclusion 5.9 *Let* $\lambda = \max\,(\lambda_1, \lambda_2)$. *Then*

$$\| f(x) - f(y) - (A(f))(x,y)(x-y) \| \le \frac{\lambda}{\Gamma\left((n+1)\,\alpha\right)} \frac{|x-y|^{(n+1)\alpha+1}}{\left((n+1)\,\alpha+1\right)},$$

(5.4.27)

$\forall\, x, y \in [a,b]$; *where* $0 < \alpha < 1,\ n \in \mathbb{N}$.

One may assume that

$$\frac{\lambda}{\Gamma\left((n+1)\,\alpha\right)} < 1.$$ (5.4.28)

Above notice that (5.4.27) is trivial when $x = y$.

Now based on (5.4.16) and (5.4.27), we can apply our numerical methods presented in this chapter to solve $f(x) = 0$.

To have $(n+1)\alpha + 1 \geq 2$, we need to take $1 > \alpha \geq \frac{1}{n+1}$, where $n \in \mathbb{N}$.

References

1. C.D. Aliprantis, K.C. Border, *Infinite Dimensional Analysis* (Springer, New York, 2006)
2. S. Amat, S. Busquier, S. Plaza, Chaotic dynamics of a third-order Newton-type method. J. Math. Anal. Appl. **366**(1), 164–174 (2010)
3. G.A. Anastassiou, Strong right fractional calculus for banach space valued functions. Revis. Proyecc. **36**(1), 149–186 (2017)
4. G.A. Anastassiou, *Principles of general fractional analysis for Banach space valued functions* (submitted for publication, 2017)
5. G.A. Anastassiou, I.K. Argyros, *Iterative convergence and applications on Banach space valued functions in right abstract fractional calculus* (submitted, 2017)
6. I.K. Argyros, F. Szidarovszky, *The Theory and Applications of Iteration Methods* (CRC Press, Boca Raton, FL, USA, 1993)
7. I.K. Argyros, A unifying local-semilocal convergence analysis and applications for two-point Newton-like methods in Banach space. J. Math. Anal. Appl. **298**, 374–397 (2004)
8. I.K. Argyros, A. Magréñan, *Iterative Methods and their Dynamics with Applications* (CRC Press, New York, 2017)
9. Bochner integral, *Encyclopedia of Mathematics*, http://www.encyclopediaofmath.org/index.php?title=Bochner_integral&oldid=38659
10. M. Edelstein, On fixed and periodic points under contractive mappings. J. Lond. Math. Soc. **37**, 74–79 (1962)
11. J.A. Ezquerro, J.M. Gutierrez, M.A. Hernandez, N. Romero, M.J. Rubio, The Newton method: from Newton to Kantorovich (Spanish). Gac. R. Soc. Mat. Esp. **13**, 53–76 (2010)
12. L.V. Kantorovich, G.P. Akilov, *Functional Analysis in Normed Spaces* (Pergamon Press, New York, 1982)
13. G.E. Ladas, V. Lakshmikantham, *Differential Equations in Abstract Spaces* (Academic Press, New York, 1972)
14. A. Magréñan, A new tool to study real dynamics: the convergence plane. Appl. Math. Comput. **248**, 215–224 (2014)
15. J. Mikusinski, *The Bochner Integral* (Academic Press, New York, 1978)
16. F.A. Potra, V. Pták, *Nondiscrete Induction and Iterative Processes* (Pitman Publication, London, 1984)
17. P.D. Proinov, New general convergence theory for iterative processes and its applications to Newton-Kantorovich type theorems. J. Complex. **26**, 3–42 (2010)
18. G.E. Shilov, *Elementary Functional Analysis* (Dover Publications Inc., New York, 1996)

Chapter 6
Algorithmic Convergence in Abstract g-Fractional Calculus

The novelty of this chapter is the design of suitable algorithms for solving equations on Banach spaces. Some applications of the semi-local convergence are suggested including Banach space valued functions of fractional calculus, where all integrals are of Bochner-type. It follows [6].

6.1 Introduction

Sections 6.1–6.2 are prerequisites for Sect. 6.3.

Let B_1, B_2 stand for Banach spaces and let Ω stand for an open subset of B_1. Let also $U(z, \rho) := \{u \in B_1 : \|u - z\| < \rho\}$ and let $\overline{U}(z, \rho)$ stand for the closure of $U(z, \rho)$.

Many problems in Computational Sciences, Engineering, Mathematical Chemistry, Mathematical Physics, Mathematical Economics and other disciplines can be brought in a form like

$$F(x) = 0 \tag{6.1.1}$$

using Mathematical Modeling [1–18], where $F : \Omega \to B_2$ is a continuous operator. The solution x^* of equation (6.1.1) is sought in closed form, but this is attainable only in special cases. That explains why most solution methods for such equations are usually iterative. There is a plethora of iterative methods for solving equation (6.1.1), more the [2, 7, 8, 10–14, 16, 17].

Newton's method [7, 8, 12, 16, 17]:

$$x_{n+1} = x_n - F'(x_n)^{-1} F(x_n). \tag{6.1.2}$$

Secant method:

$$x_{n+1} = x_n - [x_{n-1}, x_n; F]^{-1} F(x_n), \tag{6.1.3}$$

© Springer International Publishing AG 2018
G. A. Anastassiou and I. K. Argyros, *Functional Numerical Methods: Applications to Abstract Fractional Calculus*, Studies in Systems, Decision and Control 130, https://doi.org/10.1007/978-3-319-69526-6_6

where $[\cdot, \cdot; F]$ denotes a divided difference of order one on $\Omega \times \Omega$ [8, 16, 17].

Newton-like method:

$$x_{n+1} = x_n - E_n^{-1} F(x_n), \tag{6.1.4}$$

where $E_n = E(F)(x_n)$ and $E : \Omega \to \mathcal{L}(B_1, B_2)$ the space of bounded linear operators from B_1 into B_2. Other methods can be found in [8, 12, 16, 17] and the references therein.

In the present study we consider the new method defined for each $n = 0, 1, 2, \ldots$ by

$$x_{n+1} = G(x_n)$$

$$G(x_{n+1}) = G(x_n) - A_n^{-1} F(x_n), \tag{6.1.5}$$

where $x_0 \in \Omega$ is an initial point, $G : B_3 \to \Omega$ (B_3 a Banach space), $A_n = A(F)(x_{n+1}, x_n) = A(x_{n+1}, x_n)$ and $A : \Omega \times \Omega \to \mathcal{L}(B_1, B_2)$. Method (6.1.5) generates a sequence which we shall show converges to x^* under some Lipschitz-type conditions (to be precised in Sect. 6.2). Although method (6.1.5) (and Sect. 6.2) is of independent interest, it is nevertheless designed especially to be used in g-Abstract Fractional Calculus (to be precised in Sect. 6.3). As far as we know such iterative methods have not yet appeared in connection to solve equations in Abstract Fractional Calculus.

In this chapter we present the semi-local convergence of method (6.1.5) in Sect. 6.2. Some applications to Abstract g-Fractional Calculus are suggested in Sect. 6.3 on a certain Banach space valued functions, where all the integrals are of Bochner-type [9, 15].

6.2 Semi-local Convergence Analysis

We present the semi-local convergence analysis of method (6.1.5) using conditions (M):

(m_1) $F : \Omega \subset B_1 \to B_2$ is continuous, $G : B_3 \to \Omega$ is continuous and $A(x, y) \in \mathcal{L}(B_1, B_2)$ for each $(x, y) \in \Omega \times \Omega$.

(m_2) There exist $\beta > 0$ and $\Omega_0 \subset B_1$ such that $A(x, y)^{-1} \in \mathcal{L}(B_2, B_1)$ for each $(x, y) \in \Omega_0 \times \Omega_0$ and

$$\left\| A(x, y)^{-1} \right\| \leq \beta^{-1}.$$

Set $\Omega_1 = \Omega \cap \Omega_0$.

(m_3) There exists a continuous and nondecreasing function $\psi : [0, +\infty)^3 \to [0, +\infty)$ such that for each $x, y \in \Omega_1$

$$\| F(x) - F(y) - A(x, y)(G(x) - G(y)) \| \leq$$

$$\beta \psi(\|x - y\|, \|x - x_0\|, \|y - x_0\|) \|G(x) - G(y)\|.$$

(m_4) There exists a continuous and nondecreasing function $\psi_0 : [0, +\infty) \to [0, +\infty)$ such that for each $x \in \Omega_1$

$$\|G(x) - G(x_0)\| \leq \psi_0(\|x - x_0\|)\|x - x_0\|.$$

(m_5) For $x_0 \in \Omega_0$ and $x_1 = G(x_0) \in \Omega_0$ there exists $\eta \geq 0$ such that

$$\|A(x_1, x_0)^{-1} F(x_0)\| \leq \eta.$$

(m_6) There exists $s > 0$ such that

$$\psi(\eta, s, s) < 1,$$

$$\psi_0(s) < 1$$

and

$$\|G(x_0) - x_0\| \leq s \leq \frac{\eta}{1 - q_0},$$

where $q_0 = \psi(\eta, s, s)$.

(m_7) $\overline{U}(x_0, s) \subset \Omega$.

Next, we present the semi-local convergence analysis for method (6.1.5) using the conditions (M) and the preceding notation.

Theorem 6.1 *Assume that the conditions (M) hold. Then, sequence $\{x_n\}$ generated by method (6.1.5) starting at $x_0 \in \Omega$ is well defined in $U(x_0, s)$, remains in $U(x_0, s)$ for each $n = 0, 1, 2, ...$ and converges to a solution $x^* \in \overline{U}(x_0, s)$ of equation $F(x) = 0$. The limit point x^* is the unique solution of equation $F(x) = 0$ in $\overline{U}(x_0, s)$.*

Proof By the definition of s and (m_5), we have $x_1 \in U(x_0, s)$. The proof is based on mathematical induction on k. Suppose that $\|x_k - x_{k-1}\| \leq q_0^{k-1}\eta$ and $\|x_k - x_0\| \leq s$.

We get by (6.1.5), (m_2) − (m_5) in turn that

$$\|G(x_{k+1}) - G(x_k)\| = \|A_k^{-1} F(x_k)\| =$$

$$\|A_k^{-1}(F(x_k) - F(x_{k-1}) - A_{k-1}(G(x_k) - G(x_{k-1})))\|$$

$$\leq \|A_k^{-1}\| \|F(x_k) - F(x_{k-1}) - A_{k-1}(G(x_k) - G(x_{k-1}))\| \leq$$

$$\beta^{-1}\beta\psi(\|x_k - x_{k-1}\|, \|x_{k-1} - x_0\|, \|y_k - x_0\|)\|G(x_k) - G(x_{k-1})\| \leq$$

$$\psi(\eta, s, s)\|G(x_k) - G(x_{k-1})\| = q_0\|G(x_k) - G(x_{k-1})\| \leq q_0^k\|x_1 - x_0\| \leq q_0^k\eta$$
$$(6.2.1)$$

and by (m_6)

$$\|x_{k+1} - x_0\| = \|G\,(x_k) - x_0\| \le \|G\,(x_k) - G\,(x_0)\| + \|G\,(x_0) - x_0\|$$

$$\le \psi_0\,(\|x_k - x_0\|)\,\|x_k - x_0\| + \|G\,(x_0) - x_0\|$$

$$\le \psi_0\,(s)\,s + \|G\,(x_0) - x_0\| \le s.$$

The induction is completed. Moreover, we have by (6.2.1) that for $m = 0, 1, 2, \ldots$

$$\|x_{k+m} - x_k\| \le \frac{1 - q_0^m}{1 - q_0} q_0^k \eta.$$

It follows from the preceding inequation that sequence $\{G\,(x_k)\}$ is complete in a Banach space B_1 and as such it converges to some $x^* \in \overline{U}\,(x_0, s)$ (since $\overline{U}\,(x_0, s)$ is a closed ball). By letting $k \to +\infty$ in (6.2.1) we get $F\,(x^*) = 0$. We also get by (6.1.5) that $G\,(x^*) = x^*$. To show the uniqueness part, let $x^{**} \in U\,(x_0, s)$ be a solution of equation $F\,(x) = 0$ and $G\,(x^{**}) = x^{**}$. By using (6.1.5), we obtain in turn that

$$\left\|x^{**} - G\,(x_{k+1})\right\| = \left\|x^{**} - G\,(x_k) + A_k^{-1} F\,(x_k) - A_k^{-1} F\,(x^{**})\right\| \le$$

$$\left\|A_k^{-1}\right\| \left\|F\,(x^{**}) - F\,(x_k) - A_k\,(G\,(x^{**}) - G\,(x_k))\right\| \le$$

$$\beta^{-1}\beta\psi_0\,(\left\|x^{**} - x_k\right\|, \|x_{k+1} - x_0\|, \|x_k - x_0\|)\,\left\|G\,(x^{**}) - G\,(x_k)\right\| \le$$

$$q_0 \left\|G\,(x^{**}) - G\,(x_k)\right\| \le q_0^{k+1} \left\|x^{**} - x_0\right\|,$$

so $\lim\limits_{k \to +\infty} x_k = x^{**}$. We have shown that $\lim\limits_{k \to +\infty} x_k = x^*$, so $x^* = x^{**}$. ∎

Remark 6.2 (1) Condition (m_2) can become part of condition (m_3) by considering
$(m_3)'$ There exists a continuous and nondecreasing function $\varphi : [0, +\infty)^3 \to [0, +\infty)$ such that for each $x, y \in \Omega_1$

$$\left\|A\,(x, y)^{-1}\,[F\,(x) - F\,(y) - A\,(x, y)\,(G\,(x) - G\,(y))]\right\| \le$$

$$\varphi\,(\|x - y\|, \|x - x_0\|, \|y - x_0\|)\,\|G\,(x) - G\,(y)\|.$$

Notice that

$$\varphi\,(u_1, u_2, u_3) \le \psi\,(u_1, u_2, u_3)$$

for each $u_1 \ge 0$, $u_2 \ge 0$ and $u_3 \ge 0$. Similarly, a function φ_1 can replace ψ_1 for the uniqueness of the solution part. These replacements are of Mysovskii-type [7, 12, 16] and influence the weaking of the convergence criterion in (m_6), error bounds and the precision of s.

(2) Suppose that there exist $\beta > 0$, $\beta_1 > 0$ and $L \in \mathcal{L}(B_1, B_2)$ with $L^{-1} \in \mathcal{L}(B_2, B_1)$ such that

$$\left\| L^{-1} \right\| \leq \beta^{-1}$$

$$\| A(x, y) - L \| \leq \beta_1$$

and

$$\beta_2 := \beta^{-1} \beta_1 < 1.$$

Then, it follows from the Banach lemma on invertible operators [12], and

$$\left\| L^{-1} \right\| \| A(x, y) - L \| \leq \beta^{-1} \beta_1 = \beta_2 < 1$$

that $A(x, y)^{-1} \in \mathcal{L}(B_2, B_1)$. Let $\beta = \frac{\beta^{-1}}{1 - \beta_2}$. Then, under these replacements, condition (m_2) is implied, therefore it can be dropped from the conditions (M).

Remark 6.3 Section 6.2 has an interest independent of Sect. 6.3. It is worth noticing that the results especially of Theorem 6.1 can apply in Abstract g-Fractional Calculus as illustrated in Sect. 6.3. By specializing function ψ, we can apply the results of say Theorem 6.1 in the examples suggested in Sect. 6.3. In particular for (6.3.4), we choose for $u_1 \geq 0$, $u_2 \geq 0$, $u_3 \geq 0$

$$\psi(u_1, u_2, u_3) = \frac{\lambda \mu_1^{(n+1)\alpha}}{\beta \Gamma((n+1)\alpha)((n+1)\alpha + 1)},$$

if $|g(x) - g(y)| \leq \mu_1$ for each $x, y \in [a, b]$;

$$\psi(u_1, u_2, u_3) = \frac{\lambda \mu_2^{(n+1)\alpha}}{\beta \Gamma((n+1)\alpha)((n+1)\alpha + 1)},$$

if $|g(x) - g(y)| \leq \xi_2 \|x - y\|$ for each $x, y \in [a, b]$ and $\mu_2 = \xi_2 |b - a|$;

$$\psi(u_1, u_2, u_3) = \frac{\lambda \mu_3^{(n+1)\alpha}}{\beta \Gamma((n+1)\alpha)((n+1)\alpha + 1)},$$

if $|g(x)| \leq \xi_3$ for each $x, y \in [a, b]$ and $\mu_3 = 2\xi_3$.

Other choices of function ψ are also possible.

Notice that with these choices of function ψ and $f = F$ and $g = G$, crucial condition (m_3) is satisfied, which justifies our definition of method (6.1.5). We can provide similar choices for the other examples of Sect. 6.3.

6.3   Applications to X-valued Modified g-Fractional Calculus

Here we deal with Banach space $(X, \|\cdot\|)$ valued functions f of real domain $[a, b]$. All integrals here are of Bochner-type, see [15]. The derivatives of f are defined similarly to numerical ones, see [18], pp. 83–86 and p. 93.

Let $0 < \alpha \leq 1, m = \lceil \alpha \rceil = 1$ ($\lceil \cdot \rceil$ ceiling of number), g is strictly increasing and $g \in C^1([a, b])$, $g^{-1} \in C([g(a), g(b)])$. Assume that $f \in C^1([a, b], X)$. In both backgrounds here we follow [5].

(I) The X-valued right generalized g-fractional derivative of f of order α is defined as follows:

$$\left(D^\alpha_{b-;g} f \right)(x) := \frac{-1}{\Gamma(1-\alpha)} \int_x^b (g(t) - g(x))^{-\alpha} g'(t) \left(f \circ g^{-1} \right)'(g(t)) \, dt,$$
(6.3.1)

$a \leq x \leq b$.

If $0 < \alpha < 1$, then $\left(D^\alpha_{b-;g} f \right) \in C([a, b], X)$, see [4].

Also we define

$$\left(D^1_{b-;g} f \right)(x) := -\left(\left(f \circ g^{-1} \right)' \circ g \right)(x),$$
(6.3.2)

$$\left(D^0_{b-;g} f \right)(x) := f(x), \ \forall \, x \in [a, b].$$

When $g = id$, then

$$D^\alpha_{b-;g} f(x) = D^\alpha_{b-;id} f(x) = D^\alpha_{b-} f(x),$$
(6.3.3)

the usual X-valued right Caputo fractional derivative, see [3].

Denote by

$$D^{n\alpha}_{b-;g} := D^\alpha_{b-;g} D^\alpha_{b-;g} ... D^\alpha_{b-;g} \ (n \text{ times}), n \in \mathbb{N}.$$
(6.3.4)

We consider the X-valued right generalized fractional Riemann-Liouville integral

$$\left(I^\alpha_{b-;g} f \right)(x) = \frac{1}{\Gamma(\alpha)} \int_x^b (g(t) - g(x))^{\alpha-1} g'(t) f(t) \, dt, \ a \leq x \leq b. \quad (6.3.5)$$

Also denote by

$$I^{n\alpha}_{b-;g} := I^\alpha_{b-;g} I^\alpha_{b-;g} ... I^\alpha_{b-;g} \ (n \text{ times}).$$
(6.3.6)

We will be using the following X-valued modified g-right generalized Taylor's formula

Theorem 6.4 ([5]) *Let* $f \in C^1([a, b], X)$, $g \in C^1([a, b])$, *strictly increasing, such that* $g^{-1} \in C^1([g(a), g(b)])$. *Suppose that* $F_k := D^{k\alpha}_{b-;g} f, k = 1, ..., n$, *fulfill* $F_k \in C^1([a, b], X)$, *and* $F_{n+1} \in C([a, b], X)$, *where* $0 < \alpha \leq 1, n \in \mathbb{N}$. *Then*

$$f(x) - f(b) = \sum_{i=1}^{n} \frac{(g(b) - g(x))^{i\alpha}}{\Gamma(i\alpha + 1)} \left(D_{b-;g}^{i\alpha}f\right)(b) + \tag{6.3.7}$$

$$\frac{1}{\Gamma((n+1)\alpha)} \int_x^b (g(t) - g(x))^{(n+1)\alpha - 1} g'(t) \left(D_{b-;g}^{(n+1)\alpha}f\right)(t)\,dt,$$

$\forall\, x \in [a, b]$.

Here we are going to operate more generally. We consider $f \in C^1([a, b], X)$. We define the following X-valued right generalized g-fractional derivative:

$$\left(D_{y-;g}^{\alpha}f\right)(x) := \frac{-1}{\Gamma(1-\alpha)} \int_x^y (g(t) - g(x))^{-\alpha} g'(t) \left(f \circ g^{-1}\right)'(g(t))\,dt,$$
$$\tag{6.3.8}$$

all $a \le x \le y;\ y \in [a, b]$,

$$\left(D_{y-;g}^{1}f\right)(x) := -\left(\left(f \circ g^{-1}\right)' \circ g\right)(x),\ \forall\, x \in [a, b]. \tag{6.3.9}$$

Similarly we define:

$$\left(D_{x-;g}^{\alpha}f\right)(y) := \frac{-1}{\Gamma(1-\alpha)} \int_y^x (g(t) - g(y))^{-\alpha} g'(t) \left(f \circ g^{-1}\right)'(g(t))\,dt,$$
$$\tag{6.3.10}$$

all $a \le y \le x;\ x \in [a, b]$,

$$\left(D_{x-;g}^{1}f\right)(y) := -\left(\left(f \circ g^{-1}\right)' \circ g\right)(y),\ \forall\, y \in [a, b]. \tag{6.3.11}$$

When $0 < \alpha < 1$, $D_{y-;g}^{\alpha}f$ and $D_{x-;g}^{\alpha}f$ are continuous functions on $[a, b]$, see [5]. Note here that by convention we have that

$$\left(D_{y-;g}^{\alpha}f\right)(x) = 0,\ \text{for } x > y$$
$$\text{and} \tag{6.3.12}$$
$$\left(D_{x-;g}^{\alpha}f\right)(y) = 0,\ \text{for } y > x$$

Denote by

$$F_k^y := D_{y-;g}^{k\alpha}f,\ \ F_k^x := D_{x-;g}^{k\alpha}f,\ \forall\, x, y \in [a, b]. \tag{6.3.13}$$

We assume that

$$F_k^y, F_k^x \in C^1([a, b], X),\ \text{and}\ F_{n+1}^y, F_{n+1}^x \in C([a, b], X), \tag{6.3.14}$$

$k = 1, ..., n,\ \forall\, x, y \in [a, b];\ 0 < \alpha < 1$.

We also observe that $(0 < \alpha < 1)$ ([9])

$$\left\| \left(D^{\alpha}_{b-;g} f\right)(x) \right\| \leq \frac{1}{\Gamma(1-\alpha)} \int_{x}^{b} (g(t) - g(x))^{-\alpha} g'(t) \left\| \left(f \circ g^{-1}\right)'(g(t)) \right\| dt \leq \tag{6.3.15}$$

$$\frac{\left\| \left(f \circ g^{-1}\right)' \circ g \right\|_{\infty,[a,b]}}{\Gamma(1-\alpha)} \int_{x}^{b} (g(t) - g(x))^{-\alpha} g'(t) \, dt =$$

$$\frac{\left\| \left(f \circ g^{-1}\right)' \circ g \right\|_{\infty,[a,b]}}{\Gamma(1-\alpha)} \frac{(g(b) - g(x))^{1-\alpha}}{1-\alpha} =$$

$$\frac{\left\| \left(f \circ g^{-1}\right)' \circ g \right\|_{\infty,[a,b]}}{\Gamma(2-\alpha)} (g(b) - g(x))^{1-\alpha}, \ \forall \, x \in [a,b].$$

We have proved that

$$\left\| \left(D^{\alpha}_{b-;g} f\right)(x) \right\| \leq \frac{\left\| \left(f \circ g^{-1}\right)' \circ g \right\|_{\infty,[a,b]}}{\Gamma(2-\alpha)} (g(b) - g(x))^{1-\alpha} \tag{6.3.16}$$

$$\leq \frac{\left\| \left(f \circ g^{-1}\right)' \circ g \right\|_{\infty,[a,b]}}{\Gamma(2-\alpha)} (g(b) - g(a))^{1-\alpha}, \ \forall \, x, y \in [a,b].$$

Clearly here we have

$$\left(D^{\alpha}_{b-;g} f\right)(b) = 0, \ 0 < \alpha < 1. \tag{6.3.17}$$

In particular it holds

$$\left(D^{\alpha}_{x-;g} f\right)(x) = \left(D^{\alpha}_{y-;g} f\right)(y) = 0, \ \forall \, x, y \in [a,b]; \ 0 < \alpha < 1. \tag{6.3.18}$$

By (6.3.7) we derive

$$f(x) - f(y) = \sum_{i=2}^{n} \frac{(g(y) - g(x))^{i\alpha}}{\Gamma(i\alpha + 1)} \left(D^{i\alpha}_{y-;g} f\right)(y) + \tag{6.3.19}$$

$$\frac{1}{\Gamma((n+1)\alpha)} \int_{x}^{y} (g(t) - g(x))^{(n+1)\alpha - 1} g'(t) \left(D^{(n+1)\alpha}_{y-;g} f\right)(t) \, dt,$$

$\forall \, x < y; \ x, y \in [a,b]; \ 0 < \alpha < 1$, and also it holds:

$$f(y) - f(x) = \sum_{i=2}^{n} \frac{(g(x) - g(y))^{i\alpha}}{\Gamma(i\alpha + 1)} \left(D_{x-;g}^{i\alpha} f\right)(x) + \tag{6.3.20}$$

$$\frac{1}{\Gamma((n+1)\alpha)} \int_{y}^{x} (g(t) - g(y))^{(n+1)\alpha - 1} g'(t) \left(D_{x-;g}^{(n+1)\alpha} f\right)(t) \, dt,$$

$\forall \, y < x;\, x, y \in [a, b];\, 0 < \alpha < 1$.

We define also the following X-valued linear operator

$$(A_1(f))(x, y) :=$$

$$\begin{cases} \sum_{i=2}^{n} \frac{(g(y)-g(x))^{i\alpha-1}}{\Gamma(i\alpha+1)} \left(D_{y-;g}^{i\alpha} f\right)(y) - \left(D_{y-;g}^{(n+1)\alpha} f(x)\right) \frac{(g(y)-g(x))^{(n+1)\alpha-1}}{\Gamma((n+1)\alpha+1)}, & x < y, \\ \sum_{i=2}^{n} \frac{(g(x)-g(y))^{i\alpha-1}}{\Gamma(i\alpha+1)} \left(D_{x-;g}^{i\alpha} f\right)(x) - \left(D_{x-;g}^{(n+1)\alpha} f(y)\right) \frac{(g(x)-g(y))^{(n+1)\alpha-1}}{\Gamma((n+1)\alpha+1)}, & x > y, \\ f'(x), & \text{when } x = y, \end{cases}$$
$$\tag{6.3.21}$$

$\forall \, x, y \in [a, b];\, 0 < \alpha < 1$.

We may assume that (see [13], p. 3)

$$\|(A_1(f))(x, x) - (A_1(f))(y, y)\| = \|f'(x) - f'(y)\|$$

$$= \left\|\left(f' \circ g^{-1}\right)(g(x)) - \left(f' \circ g^{-1}\right)(g(y))\right\| \le \Phi \, |g(x) - g(y)|, \tag{6.3.22}$$

$\forall \, x, y \in [a, b];$ with $\Phi > 0$.

We estimate and have

(i) case $x < y$:

$$\|f(x) - f(y) - (A_1(f))(x, y)(g(x) - g(y))\| =$$

$$\left\| \frac{1}{\Gamma((n+1)\alpha)} \int_{x}^{y} (g(t) - g(x))^{(n+1)\alpha - 1} g'(t) \left(D_{y-;g}^{(n+1)\alpha} f\right)(t) \, dt - \right.$$

$$\left. \left(D_{y-;g}^{(n+1)\alpha} f(x)\right) \frac{(g(y) - g(x))^{(n+1)\alpha}}{\Gamma((n+1)\alpha+1)} \right\| \tag{6.3.23}$$

(by [1], p. 426, Theorem 11.43)

$$= \frac{1}{\Gamma((n+1)\alpha)} \cdot$$

$$\left\| \int_{x}^{y} (g(t) - g(x))^{(n+1)\alpha - 1} g'(t) \left(\left(D_{y-;g}^{(n+1)\alpha} f\right)(t) - \left(D_{y-;g}^{(n+1)\alpha} f\right)(x)\right) dt \right\|$$

(by [9])

$$\leq \frac{1}{\Gamma\left((n+1)\alpha\right)} \cdot$$

$$\int_x^y \left(g\left(t\right) - g\left(x\right)\right)^{(n+1)\alpha-1} g'\left(t\right) \left\| \left(D_{y-;g}^{(n+1)\alpha} f\right)\left(t\right) - \left(D_{y-;g}^{(n+1)\alpha} f\right)\left(x\right) \right\| dt \quad (6.3.24)$$

(we assume that

$$\left\| \left(D_{y-;g}^{(n+1)\alpha} f\right)\left(t\right) - \left(D_{y-;g}^{(n+1)\alpha} f\right)\left(x\right) \right\| \leq \lambda_1 \left| g\left(t\right) - g\left(x\right) \right|, \qquad (6.3.25)$$

$\forall\, t, x, y \in [a, b] : y \geq t \geq x;\ \lambda_1 > 0$)

$$\leq \frac{\lambda_1}{\Gamma\left((n+1)\alpha\right)} \int_x^y \left(g\left(t\right) - g\left(x\right)\right)^{(n+1)\alpha-1} g'\left(t\right) \left(g\left(t\right) - g\left(x\right)\right) dt = \quad (6.3.26)$$

$$\frac{\lambda_1}{\Gamma\left((n+1)\alpha\right)} \int_x^y \left(g\left(t\right) - g\left(x\right)\right)^{(n+1)\alpha} g'\left(t\right) dt =$$

$$\frac{\lambda_1}{\Gamma\left((n+1)\alpha\right)} \frac{\left(g\left(y\right) - g\left(x\right)\right)^{(n+1)\alpha+1}}{\left((n+1)\alpha + 1\right)}. \qquad (6.3.27)$$

We have proved that

$$\left\| f\left(x\right) - f\left(y\right) - \left(A_1\left(f\right)\right)\left(x, y\right)\left(g\left(x\right) - g\left(y\right)\right) \right\| \leq$$

$$\frac{\lambda_1}{\Gamma\left((n+1)\alpha\right)} \frac{\left(g\left(y\right) - g\left(x\right)\right)^{(n+1)\alpha+1}}{\left((n+1)\alpha + 1\right)}, \qquad (6.3.28)$$

for any $x, y \in [a, b] : x < y;\ 0 < \alpha < 1$.

(ii) case $x > y$:

$$\left\| f\left(x\right) - f\left(y\right) - \left(A_1\left(f\right)\right)\left(x, y\right)\left(g\left(x\right) - g\left(y\right)\right) \right\| =$$

$$\left\| f\left(y\right) - f\left(x\right) - \left(A_1\left(f\right)\right)\left(x, y\right)\left(g\left(y\right) - g\left(x\right)\right) \right\| = \qquad (6.3.29)$$

$$\left\| \frac{1}{\Gamma\left((n+1)\alpha\right)} \int_y^x \left(g\left(t\right) - g\left(y\right)\right)^{(n+1)\alpha-1} g'\left(t\right) \left(D_{x-;g}^{(n+1)\alpha} f\right)\left(t\right) dt - \right.$$

$$\left. \left(D_{x-;g}^{(n+1)\alpha} f\right)\left(y\right) \frac{\left(g\left(x\right) - g\left(y\right)\right)^{(n+1)\alpha}}{\Gamma\left((n+1)\alpha + 1\right)} \right\| \qquad (6.3.30)$$

$$= \frac{1}{\Gamma\left((n+1)\alpha\right)} \cdot$$

$$\left\| \int_y^x (g(t) - g(y))^{(n+1)\alpha-1} g'(t) \left(\left(D_{x-;g}^{(n+1)\alpha} f \right)(t) - \left(D_{x-;g}^{(n+1)\alpha} f \right)(y) \right) dt \right\| \le$$

$$\frac{1}{\Gamma((n+1)\alpha)} \int_y^x (g(t) - g(y))^{(n+1)\alpha-1} g'(t) \left\| D_{x-;g}^{(n+1)\alpha} f(t) - D_{x-;g}^{(n+1)\alpha} f(y) \right\| dt$$

(we assume that

$$\left\| D_{x-;g}^{(n+1)\alpha} f(t) - D_{x-;g}^{(n+1)\alpha} f(y) \right\| \le \lambda_2 \left| g(t) - g(y) \right|, \tag{6.3.31}$$

$\forall\, t, y, x \in [a, b] : x \ge t \ge y; \lambda_2 > 0)$

$$\le \frac{\lambda_2}{\Gamma((n+1)\alpha)} \int_y^x (g(t) - g(y))^{(n+1)\alpha-1} g'(t) (g(t) - g(y)) \, dt = \tag{6.3.32}$$

$$\frac{\lambda_2}{\Gamma((n+1)\alpha)} \int_y^x (g(t) - g(y))^{(n+1)\alpha} g'(t) \, dt =$$

$$\frac{\lambda_2}{\Gamma((n+1)\alpha)} \frac{(g(x) - g(y))^{(n+1)\alpha+1}}{((n+1)\alpha+1)}.$$

We have proved that

$$\| f(x) - f(y) - (A_1(f))(x, y)(g(x) - g(y)) \| \le$$

$$\frac{\lambda_2}{\Gamma((n+1)\alpha)} \frac{(g(x) - g(y))^{(n+1)\alpha+1}}{((n+1)\alpha+1)}, \tag{6.3.33}$$

$\forall\, x, y \in [a, b] : x > y; 0 < \alpha < 1.$

Conclusion 6.5 *Set $\lambda = \max(\lambda_1, \lambda_2)$. We have proved that*

$$\| f(x) - f(y) - (A_1(f))(x, y)(g(x) - g(y)) \| \le$$

$$\frac{\lambda}{\Gamma((n+1)\alpha)} \frac{|g(x) - g(y)|^{(n+1)\alpha+1}}{((n+1)\alpha+1)}, \tag{6.3.34}$$

$\forall\, x, y \in [a, b]; 0 < \alpha < 1, n \in \mathbb{N}.$

(Notice that (6.3.34) is trivially true when $x = y$.)
One may assume that

$$\frac{\lambda}{\Gamma((n+1)\alpha)} < 1. \tag{6.3.35}$$

Now based on (6.3.22) and (6.3.34), we can apply our numerical methods presented in this chapter to solve $f(x) = 0$.

To have $(n+1)\alpha + 1 \geq 2$, we need to take $1 > \alpha \geq \frac{1}{n+1}$, where $n \in \mathbb{N}$.

Some examples of g follow:

$$
\begin{aligned}
&g(x) = e^x, \ x \in [a, b] \subset \mathbb{R}, \\
&g(x) = \sin x, \\
&g(x) = \tan x, \\
&\text{where } x \in \left[-\frac{\pi}{2} + \varepsilon, \frac{\pi}{2} - \varepsilon\right], \ \varepsilon > 0 \text{ small.}
\end{aligned}
\tag{6.3.36}
$$

(II) The X-valued left generalized g-fractional derivative of f of order α is defined as follows (see [5]):

$$
\left(D_{a+;g}^{\alpha} f\right)(x) = \frac{1}{\Gamma(1-\alpha)} \int_a^x (g(x) - g(t))^{-\alpha} g'(t) \left(f \circ g^{-1}\right)'(g(t)) \, dt,
\tag{6.3.37}
$$

$\forall \, x \in [a, b]$.

If $0 < \alpha < 1$, then $\left(D_{a+;g}^{\alpha} f\right) \in C([a, b], X)$ (see [5]).

Also, we define

$$
D_{a+;g}^1 f(x) = \left(\left(f \circ g^{-1}\right)' \circ g\right)(x),
\tag{6.3.38}
$$

$$
D_{a+;g}^0 f(x) = f(x), \ \forall \, x \in [a, b].
$$

When $g = id$, then

$$
D_{a+;g}^{\alpha} f = D_{a+;id}^{\alpha} f = D_{*a}^{\alpha} f,
$$

the usual X-valued left Caputo fractional derivative (see [4]).

Denote by

$$
D_{a+;g}^{n\alpha} := D_{a+;g}^{\alpha} D_{a+;g}^{\alpha} \ldots D_{a+;g}^{\alpha} \ (n \text{ times}), \ n \in \mathbb{N}.
\tag{6.3.39}
$$

We consider the X-valued left generalized fractional Riemann-Liouville integral (see [5])

$$
\left(I_{a+;g}^{\alpha} f\right)(x) = \frac{1}{\Gamma(\alpha)} \int_a^x (g(x) - g(t))^{\alpha-1} g'(t) f(t) \, dt, \ a \leq x \leq b.
\tag{6.3.40}
$$

Also denote by

$$
I_{a+;g}^{n\alpha} := I_{a+;g}^{\alpha} I_{a+;g}^{\alpha} \ldots I_{a+;g}^{\alpha} \ (n \text{ times}).
\tag{6.3.41}
$$

We will be using the following X-valued modified g-left generalized Taylor's formula.

Theorem 6.6 ([5]) *Let* $0 < \alpha \leq 1$, $n \in \mathbb{N}$, $f \in C^1([a, b], X)$, $g \in C^1([a, b])$, *strictly increasing, such that* $g^{-1} \in C^1([g(a), g(b)])$. *Let* $F_k := D_{a+;g}^{k\alpha} f$, $k = 1, \ldots, n$, *that fulfill* $F_k \in C^1([a, b], X)$, *and* $F_{n+1} \in C([a, b], X)$. *Then*

$$f(x) - f(a) = \sum_{i=1}^{n} \frac{(g(x) - g(a))^{i\alpha}}{\Gamma(i\alpha + 1)} \left(D_{a+;g}^{i\alpha} f\right)(a) + \tag{6.3.42}$$

$$\frac{1}{\Gamma((n+1)\alpha)} \int_a^x (g(x) - g(t))^{(n+1)\alpha - 1} g'(t) \left(D_{a+;g}^{(n+1)\alpha} f\right)(t)\, dt,$$

$\forall x \in [a, b]$.

Here we are going to operate more generally. We consider $f \in C^1([a, b], X)$. We define the following X-valued left generalized g-fractional derivative:

$$\left(D_{y+;g}^{\alpha} f\right)(x) = \frac{1}{\Gamma(1 - \alpha)} \int_y^x (g(x) - g(t))^{-\alpha} g'(t) \left(f \circ g^{-1}\right)'(g(t))\, dt, \tag{6.3.43}$$

for any $y \le x \le b$; $x, y \in [a, b]$,

$$\left(D_{y+;g}^1 f\right)(x) = \left(f \circ g^{-1}\right)'(g(x)), \ \forall x \in [a, b]. \tag{6.3.44}$$

Similarly, we define

$$\left(D_{x+;g}^{\alpha} f\right)(y) = \frac{1}{\Gamma(1 - \alpha)} \int_x^y (g(y) - g(t))^{-\alpha} g'(t) \left(f \circ g^{-1}\right)'(g(t))\, dt, \tag{6.3.45}$$

for any $x \le y \le b$; $x, y \in [a, b]$,

$$\left(D_{x+;g}^1 f\right)(y) = \left(f \circ g^{-1}\right)'(g(y)), \ \forall y \in [a, b]. \tag{6.3.46}$$

When $0 < \alpha < 1$, $D_{y+;g}^{\alpha} f$ and $D_{x+;g}^{\alpha} f$ are continuous functions on $[a, b]$, see [5]. Note here that by convention, we have that

$$\begin{aligned} \left(D_{y+;g}^{\alpha} f\right)(x) &= 0, \text{ when } x < y, \\ \text{and} & \\ \left(D_{x+;g}^{\alpha} f\right)(y) &= 0, \text{ when } y < x. \end{aligned} \tag{6.3.47}$$

Denote by

$$G_k^y := D_{y+;g}^{k\alpha} f, \ G_k^x := D_{x+;g}^{k\alpha} f, \ \forall x, y \in [a, b]. \tag{6.3.48}$$

We assume that

$$G_k^y, G_k^x \in C^1([a, b], X), \text{ and } G_{n+1}^y, G_{n+1}^x \in C([a, b], X), \tag{6.3.49}$$

$k = 1, ..., n, \ \forall x, y \in [a, b]; \ 0 < \alpha < 1$.

We also observe that $(0 < \alpha < 1)$ (by [9])

$$\left\| \left(D_{a+;g}^{\alpha} f \right)(x) \right\| \le \frac{1}{\Gamma(1-\alpha)} \int_a^x (g(x) - g(t))^{-\alpha} g'(t) \left\| \left(f \circ g^{-1} \right)'(g(t)) \right\| dt \le$$

$$\frac{\left\| \left(f \circ g^{-1} \right)' \circ g \right\|_{\infty,[a,b]}}{\Gamma(1-\alpha)} \int_a^x (g(x) - g(t))^{-\alpha} g'(t) \, dt =$$

$$\frac{\left\| \left(f \circ g^{-1} \right)' \circ g \right\|_{\infty,[a,b]}}{\Gamma(1-\alpha)} \frac{(g(x) - g(a))^{1-\alpha}}{1-\alpha} = \qquad (6.3.50)$$

$$\frac{\left\| \left(f \circ g^{-1} \right)' \circ g \right\|_{\infty,[a,b]}}{\Gamma(2-\alpha)} (g(x) - g(a))^{1-\alpha}.$$

We have proved that

$$\left\| \left(D_{a+;g}^{\alpha} f \right)(x) \right\| \le \frac{\left\| \left(f \circ g^{-1} \right)' \circ g \right\|_{\infty,[a,b]}}{\Gamma(2-\alpha)} (g(x) - g(a))^{1-\alpha}$$

$$\le \frac{\left\| \left(f \circ g^{-1} \right)' \circ g \right\|_{\infty,[a,b]}}{\Gamma(2-\alpha)} (g(b) - g(a))^{1-\alpha}, \ \forall \, x \in [a,b]. \qquad (6.3.51)$$

In particular it holds

$$\left(D_{a+;g}^{\alpha} f \right)(a) = 0, \ 0 < \alpha < 1, \qquad (6.3.52)$$

and

$$\left(D_{y+;g}^{\alpha} f \right)(y) = \left(D_{x+;g}^{\alpha} f \right)(x) = 0, \ \forall \, x, y \in [a,b]; \ 0 < \alpha < 1. \qquad (6.3.53)$$

By (6.3.42) we derive

$$f(x) - f(y) = \sum_{i=2}^{n} \frac{(g(x) - g(y))^{i\alpha}}{\Gamma(i\alpha + 1)} \left(D_{y+;g}^{i\alpha} f \right)(y) +$$

$$\frac{1}{\Gamma((n+1)\alpha)} \int_y^x (g(x) - g(t))^{(n+1)\alpha - 1} g'(t) \left(D_{y+;g}^{(n+1)\alpha} f \right)(t) \, dt, \qquad (6.3.54)$$

for any $x > y : x, y \in [a,b]; \ 0 < \alpha < 1$, also it holds

$$f(y) - f(x) = \sum_{i=2}^{n} \frac{(g(y) - g(x))^{i\alpha}}{\Gamma(i\alpha + 1)} \left(D_{x+;g}^{i\alpha} f\right)(x) + \tag{6.3.55}$$

$$\frac{1}{\Gamma((n+1)\alpha)} \int_x^y (g(y) - g(t))^{(n+1)\alpha-1} g'(t) \left(D_{x+;g}^{(n+1)\alpha} f\right)(t) \, dt,$$

for any $y > x : x, y \in [a, b]; 0 < \alpha < 1$.

We define also the following X-valued linear operator

$$(A_2(f))(x, y) :=$$

$$\begin{cases} \sum_{i=2}^{n} \frac{(g(x)-g(y))^{i\alpha-1}}{\Gamma(i\alpha+1)} \left(D_{y+;g}^{i\alpha} f\right)(y) + \left(D_{y+;g}^{(n+1)\alpha} f\right)(x) \frac{(g(x)-g(y))^{(n+1)\alpha-1}}{\Gamma((n+1)\alpha+1)}, & x > y, \\ \sum_{i=2}^{n} \frac{(g(y)-g(x))^{i\alpha-1}}{\Gamma(i\alpha+1)} \left(D_{x+;g}^{i\alpha} f\right)(x) + \left(D_{x+;g}^{(n+1)\alpha} f\right)(y) \frac{(g(y)-g(x))^{(n+1)\alpha-1}}{\Gamma((n+1)\alpha+1)}, & y > x, \\ f'(x), & \text{when } x = y, \end{cases}$$

$$\tag{6.3.56}$$

$\forall x, y \in [a, b]; 0 < \alpha < 1$.

We may assume that (see [13], p. 3)

$$\|(A_2(f))(x, x) - (A_2(f))(y, y)\| = \|f'(x) - f'(y)\| \tag{6.3.57}$$

$$\leq \Phi^* |g(x) - g(y)|, \ \forall x, y \in [a, b];$$

with $\Phi^* > 0$.

We estimate and have
(i) case of $x > y$:

$$\|f(x) - f(y) - (A_2(f))(x, y)(g(x) - g(y))\| = \tag{6.3.58}$$

$$\left\| \frac{1}{\Gamma((n+1)\alpha)} \int_y^x (g(x) - g(t))^{(n+1)\alpha-1} g'(t) \left(D_{y+;g}^{(n+1)\alpha} f\right)(t) \, dt - \right.$$

$$\left. \left(D_{y+;g}^{(n+1)\alpha} f\right)(x) \frac{(g(x) - g(y))^{(n+1)\alpha}}{\Gamma((n+1)\alpha+1)} \right\|$$

(by [1], p. 426, Theorem 11.43)

$$= \frac{1}{\Gamma((n+1)\alpha)} \cdot$$

$$\left\| \int_y^x (g(x) - g(tx))^{(n+1)\alpha-1} g'(t) \left(\left(D_{y+;g}^{(n+1)\alpha} f\right)(t) - \left(D_{y+;g}^{(n+1)\alpha} f\right)(x)\right) dt \right\|$$

$$\tag{6.3.59}$$

(by [9])

$$\leq \frac{1}{\Gamma((n+1)\alpha)} \cdot$$

$$\int_y^x (g(x) - g(t))^{(n+1)\alpha-1} g'(t) \left\| \left(D_{y+;g}^{(n+1)\alpha} f\right)(t) - \left(D_{y+;g}^{(n+1)\alpha} f\right)(x) \right\| dt$$

(we assume here that

$$\left\| \left(D_{y+;g}^{(n+1)\alpha} f\right)(t) - \left(D_{y+;g}^{(n+1)\alpha} f\right)(x) \right\| \leq \rho_1 |g(t) - g(x)|, \tag{6.3.60}$$

$\forall \, t, x, y \in [a, b] : x \geq t \geq y; \, \rho_1 > 0$)

$$\leq \frac{\rho_1}{\Gamma((n+1)\alpha)} \int_y^x (g(x) - g(t))^{(n+1)\alpha-1} g'(t) (g(x) - g(t)) \, dt =$$

$$\frac{\rho_1}{\Gamma((n+1)\alpha)} \int_y^x (g(x) - g(t))^{(n+1)\alpha} g'(t) \, dt =$$

$$\frac{\rho_1}{\Gamma((n+1)\alpha)} \frac{(g(x) - g(y))^{(n+1)\alpha+1}}{((n+1)\alpha+1)}. \tag{6.3.61}$$

We have proved that

$$\|f(x) - f(y) - (A_2(f))(x, y)(g(x) - g(y))\| \leq$$

$$\frac{\rho_1}{\Gamma((n+1)\alpha)} \frac{(g(x) - g(y))^{(n+1)\alpha+1}}{((n+1)\alpha+1)}, \tag{6.3.62}$$

$\forall \, x, y \in [a, b] : x > y; \, 0 < \alpha < 1$.
 (ii) case of $y > x$:

$$\|f(x) - f(y) - (A_2(f))(x, y)(g(x) - g(y))\| = \tag{6.3.63}$$

$$\|f(y) - f(x) - (A_2(f))(x, y)(g(y) - g(x))\| =$$

$$\left\| \frac{1}{\Gamma((n+1)\alpha)} \int_x^y (g(y) - g(t))^{(n+1)\alpha-1} g'(t) \left(D_{x+;g}^{(n+1)\alpha} f\right)(t) \, dt - \right.$$

$$\left. \left(D_{x+;g}^{(n+1)\alpha} f\right)(y) \frac{(g(y) - g(x))^{(n+1)\alpha}}{\Gamma((n+1)\alpha+1)} \right\|$$

$$= \frac{1}{\Gamma((n+1)\alpha)} \cdot$$

$$\left\| \int_x^y (g(y) - g(t))^{(n+1)\alpha - 1} g'(t) \left(\left(D_{x+;g}^{(n+1)\alpha} f \right)(t) - \left(D_{x+;g}^{(n+1)\alpha} f \right)(y) \right) dt \right\|$$

$$\leq \frac{1}{\Gamma((n+1)\alpha)} \cdot$$

$$\int_x^y (g(y) - g(t))^{(n+1)\alpha - 1} g'(t) \left\| \left(D_{x+;g}^{(n+1)\alpha} f \right)(t) - \left(D_{x+;g}^{(n+1)\alpha} f \right)(y) \right\| dt \quad (6.3.64)$$

(we assume here that

$$\left\| \left(D_{x+;g}^{(n+1)\alpha} f \right)(t) - \left(D_{x+;g}^{(n+1)\alpha} f \right)(y) \right\| \leq \rho_2 |g(t) - g(y)|, \qquad (6.3.65)$$

$\forall \, t, y, x \in [a, b] : y \geq t \geq x; \rho_2 > 0$)

$$\leq \frac{\rho_2}{\Gamma((n+1)\alpha)} \int_x^y (g(y) - g(t))^{(n+1)\alpha - 1} g'(t)(g(y) - g(t)) \, dt =$$

$$\frac{\rho_2}{\Gamma((n+1)\alpha)} \frac{(g(y) - g(x))^{(n+1)\alpha + 1}}{((n+1)\alpha + 1)}. \qquad (6.3.66)$$

We have proved that

$$\| f(x) - f(y) - (A_2(f))(x, y)(g(x) - g(y)) \| \leq$$

$$\frac{\rho_2}{\Gamma((n+1)\alpha)} \frac{(g(y) - g(x))^{(n+1)\alpha + 1}}{((n+1)\alpha + 1)},$$

$\forall \, x, y \in [a, b] : y > x; 0 < \alpha < 1.$

Conclusion 6.7 *Set $\rho = \max(\rho_1, \rho_2)$. Then*

$$\| f(x) - f(y) - (A_2(f))(x, y)(g(x) - g(y)) \| \leq$$

$$\frac{\rho}{\Gamma((n+1)\alpha)} \frac{|g(x) - g(y)|^{(n+1)\alpha + 1}}{((n+1)\alpha + 1)}, \qquad (6.3.67)$$

$\forall \, x, y \in [a, b]; 0 < \alpha < 1.$

(Notice (6.3.67) is trivially true when $x = y$.)
One may assume that

$$\frac{\rho}{\Gamma((n+1)\alpha)} < 1. \qquad (6.3.68)$$

Now based on (6.3.57) and (6.3.67), we can apply our numerical methods presented in this chapter to solve $f(x) = 0$.

References

1. C.D. Aliprantis, K.C. Border, *Infinite Dimensional Analysis* (Springer, New York, 2006)
2. S. Amat, S. Busquier, S. Plaza, Chaotic dynamics of a third-order Newton-type method. J. Math. Anal. Applic. **366**(1), 164–174 (2010)
3. G.A. Anastassiou, Strong right fractional calculus for banach space valued functions. Rev. Proyecc. **36**(1), 149–186 (2017)
4. G.A. Anastassiou, *A strong Fractional Calculus Theory for Banach space valued functions*, Nonlinear Functional Analysis and Applications (Korea) (2017). accepted for publication
5. G.A. Anastassiou, *Principles of general fractional analysis for Banach space valued functions* (2017). submitted for publication
6. G.A. Anastassiou, I.K. Argyros, Algorithmic convergence on Banach space valued functions in abstract g-fractional calculus. Prog. Fract. Differ. Appl. (2017). accepted
7. I.K. Argyros, A unifying local-semilocal convergence analysis and applications for two-point Newton-like methods in Banach space. J. Math. Anal. Appl. **298**, 374–397 (2004)
8. I.K. Argyros, A. Magréñan, *Iterative methods and their dynamics with applications* (CRC Press, New York, 2017)
9. Bochner integral, Encyclopedia of Mathematics, http://www.encyclopediaofmath.org/index.php?title=Bochner_integral&oldid=38659
10. M. Edelstein, On fixed and periodic points under contractive mappings. J. London Math. Soc. **37**, 74–79 (1962)
11. J.A. Ezquerro, J.M. Gutierrez, M.A. Hernandez, N. Romero, M.J. Rubio, The Newton method: From Newton to Kantorovich (Spanish). Gac. R. Soc. Mat. Esp. **13**, 53–76 (2010)
12. L.V. Kantorovich, G.P. Akilov, *Functional Analysis in Normed Spaces* (Pergamon Press, New York, 1982)
13. G.E. Ladas, V. Lakshmikantham, *Differential equations in abstract spaces* (Academic Press, New York, London, 1972)
14. A. Magréñan, A new tool to study real dynamics: the convergence plane. Appl. Math. Comput. **248**, 215–224 (2014)
15. J. Mikusinski, *The Bochner Integral* (Academic Press, New York, 1978)
16. F.A. Potra, V. Ptăk, *Nondiscrete Induction and Iterative Processes* (Pitman Publishing, London, 1984)
17. P.D. Proinov, New general convergence theory for iterative processes and its applications to Newton-Kantorovich type theorems. J. Complex. **26**, 3–42 (2010)
18. G.E. Shilov, *Elementary Functional Analysis* (Dover Publications Inc, New York, 1996)

Chapter 7
Iterative Procedures for Solving Equations in Abstract Fractional Calculus

The objective in this study is to use generalized iterative procedures in order to approximate solutions of an equation on a Banach space setting. In particular, we present a semi-local convergence analysis for these methods. Some applications are suggested including Banach space valued functions of fractional calculus, where all integrals are of Bochner-type. It follows [5].

7.1 Introduction

Sections 7.1–7.3 are prerequisites for Sect. 7.4.

Let B_1, B_2 stand for Banach spaces and let Ω stand for an open subset of B_1. Let also $U(z, \rho) := \{u \in B_1 : \|u - z\| < \rho\}$ and let $\overline{U}(z, \rho)$ stand for the closure of $U(z, \rho)$.

Numerous problems in Computational Sciences, Engineering, Mathematical Chemistry, Mathematical Physics, Mathematical Economics and other disciplines can be brought in a form like

$$F(x) = 0 \qquad (7.1.1)$$

using Mathematical Modeling [1–17], where $F : \Omega \to B_2$ is a continuous operator. The solution x^* of Eq. (7.1.1) is sought in closed form, but this is attainable only in special cases. That explains why most solution methods for such equations are usually iterative. There is a plethora of iterative methods for solving Eq. (7.1.1). We can divide these methods in two categories.

Explicit Methods [6, 7, 11, 15, 16]: Newton's method

$$x_{n+1} = x_n - F'(x_n)^{-1} F(x_n). \qquad (7.1.2)$$

Secant method:

$$x_{n+1} = x_n - [x_{n-1}, x_n; F]^{-1} F(x_n), \qquad (7.1.3)$$

© Springer International Publishing AG 2018
G. A. Anastassiou and I. K. Argyros, *Functional Numerical Methods: Applications to Abstract Fractional Calculus*, Studies in Systems, Decision and Control 130, https://doi.org/10.1007/978-3-319-69526-6_7

where $[\cdot, \cdot; F]$ denotes a divided difference of order one on $\Omega \times \Omega$ [7, 14, 15].

Newton-like method:

$$x_{n+1} = x_n - E_n^{-1} F(x_n), \qquad (7.1.4)$$

where $E_n = E(F)(x_n)$ and $E : \Omega \rightarrow \mathcal{L}(B_1, B_2)$ the space of bounded linear operators from B_1 into B_2. Other explicit methods can be found in [7, 11, 14, 15] and the references there in.

Implicit Methods [6, 9, 11, 16]:

$$F(x_n) + A_n (x_{n+1} - x_n) = 0 \qquad (7.1.5)$$

$$x_{n+1} = x_n - A_n^{-1} F(x_n), \qquad (7.1.6)$$

where $A_n = A(x_{n+1}, x_n) = A(F)(x_{n+1}, x_n)$ and $A : \Omega \times \Omega \rightarrow \mathcal{L}(B_1, B_2)$. We also denote $A(F)(u, v) = A(F)(u) = A(u)$, if $u = v$ for each $u, v \in \Omega$.

There is a plethora on local as well as semi-local convergence results for explicit methods [1–7, 9–16]. However, the research on the convergence of implicit methods has received little attention. Authors, usually consider the fixed point problem

$$P_z(x) = x, \qquad (7.1.7)$$

where

$$P_z(x) = x + F(z) + A(x, z)(x - z) \qquad (7.1.8)$$

or

$$P_z(x) = z - A(x, z)^{-1} F(z) \qquad (7.1.9)$$

for methods (7.1.5) and (7.1.6), respectively, where $z \in \Omega$ is given. If P is a contraction operator mapping a closed set into itself, then according to the contraction mapping principle [11, 15, 16], P_z has a fixed point x_z^* which can be found using the method of successive substitutions or Picard's method [16] defined for each fixed n by

$$y_{k+1,n} = P_{x_n}(y_{k,n}), \quad y_{0,n} = x_n, \quad x_{n+1} = \lim_{k \to +\infty} y_{k,n}. \qquad (7.1.10)$$

Let us also consider the analogous explicit methods

$$F(x_n) + A(x_n, x_n)(x_{n+1} - x_n) = 0 \qquad (7.1.11)$$

$$x_{n+1} = x_n - A(x_n, x_n)^{-1} F(x_n) \qquad (7.1.12)$$

$$F(x_n) + A(x_n, x_{n-1})(x_{n+1} - x_n) = 0 \qquad (7.1.13)$$

and

$$x_{n+1} = x_n - A(x_n, x_{n-1})^{-1} F(x_n). \qquad (7.1.14)$$

In this chapter in Sect. 7.2, we study the semi-local convergence of method (7.1.5) and method (7.1.6). Section 7.3 contains the semi-local convergence of method (7.1.11), method (7.1.12), method (7.1.13) and method (7.1.14). Some applications to Abstract Fractional Calculus are suggested in Sect. 7.4 on a certain Banach space valued functions, where all the integrals are of Bochner-type [8, 14].

7.2 Semi-local Convergence for Implicit Methods

The semi-local convergence analysis of method (7.1.6) that follows is based on the conditions (H):

(h_1) $F : \Omega \subset B_1 \to B_2$ is continuous and $A(F)(x, y) \in \mathcal{L}(B_1, B_2)$ for each $(x, y) \in \Omega \times \Omega$.

(h_2) There exist $l > 0$ and $\Omega_0 \subset B_1$ such that $A(F)(x, y)^{-1} \in \mathcal{L}(B_2, B_1)$ for each $(x, y) \in \Omega_0 \times \Omega_0$ and

$$\left\| A(F)(x, y)^{-1} \right\| \leq l^{-1}.$$

Set $\Omega_1 = \Omega \cap \Omega_0$.

(h_3) There exist real numbers $\alpha_1, \alpha_2, \alpha_3$ satisfying

$$0 \leq \alpha_2 \leq \alpha_1 \text{ and } 0 \leq \alpha_3 < 1$$

such that for each $x, y \in \Omega_1$

$$\| F(x) - F(y) - A(F)(x, y)(x - y) \| \leq$$

$$l \left(\frac{\alpha_1}{2} \|x - y\| + \alpha_2 \|y - x_0\| + \alpha_3 \right) \|x - y\| .$$

(h_4) For each $x \in \Omega_0$ there exists $y \in \Omega_0$ such that

$$y = x - A(y, x)^{-1} F(x) .$$

(h_5) For $x_0 \in \Omega_0$ and $x_1 \in \Omega_0$ satisfying (h_4) there exists $\eta \geq 0$ such that

$$\left\| A(F)(x_1, x_0)^{-1} F(x_0) \right\| \leq \eta.$$

(h_6) $h := \alpha_1 \eta \leq \frac{1}{2}(1 - \alpha_3)^2 .$

and

(h_7) $\overline{U}(x_0, t^*) \subset \Omega_0$, where

$$t^* = \begin{cases} \frac{1-\alpha_3-\sqrt{(1-\alpha_3)^2-2h}}{\alpha_1}, & \alpha_1 \neq 0 \\ \frac{1}{1-\alpha_3}\eta, & \alpha_1 = 0. \end{cases}$$

Then, we can show the following semi-local convergence result for method (7.1.6) under the preceding notation and conditions (H).

Theorem 7.1 *Suppose that the conditions (H) are satisfied. Then, sequence $\{x_n\}$ generated by method (7.1.6) starting at $x_0 \in \Omega$ is well defined in $U(x_0, t^*)$, remains in $U(x_0, t^*)$ for each $n = 0, 1, 2, \dots$ and converges to a solution $x^* \in \overline{U}(x_0, t^*)$ of equation $F(x) = 0$. Moreover, provided that (h_3) holds with $A(F)(z,y)$ replacing $A(F)(x,y)$ for each $z \in \Omega_1$, if $\alpha_1 \neq 0$, the equation $F(x) = 0$ has a unique solution x^* in $\widetilde{U}$, where*

$$\widetilde{U} = \begin{cases} \overline{U}(x_0, t^*) \cap \Omega_0, & \text{if } h = \frac{1}{2}(1 - \alpha_3)^2 \\ U(x_0, t^{**}) \cap \Omega_0, & \text{if } h < \frac{1}{2}(1 - \alpha_3)^2 \end{cases}$$

and, if $\alpha_1 = 0$, the solution x^* is unique in $\overline{U}\left(x_0, \frac{\eta}{1-\alpha_3}\right)$, where $t^{**} = \frac{1-\alpha_3+\sqrt{(1-\alpha_3)^2-2h}}{\alpha_1}$.

Proof Case $\alpha_1 \neq 0$. Let us define scalar function g on $\mathbb{R}$ by $g(t) = \frac{\alpha_1}{2}t^2 - (1 - \alpha_3)t + \eta$ and majorizing sequence $\{t_n\}$ by

$$t_0 = 0, \quad t_k = t_{k-1} + g(t_{k-1}) \text{ for each } k = 1, 2, \dots. \tag{7.2.1}$$

It follows from (h_6) that function g has two positive roots t^* and t^{**}, $t^* \leq t^{**}$, and $t_k \leq t_{k+1}$. That is, sequence $\{t_k\}$ converges to t^*.

(a) Using mathematical induction on k, we shall show that

$$\|x_{k+1} - x_k\| \leq t_{k+1} - t_k. \tag{7.2.2}$$

Estimate (7.2.2) holds for $k = 0$ by (h_5) and (7.2.1), since $\|x_1 - x_0\| \leq \eta = t_1 - t_0$. Suppose that for $1 \leq m \leq k$

$$\|x_m - x_{m-1}\| \leq t_m - t_{m-1}. \tag{7.2.3}$$

Them, we get $\|x_k - x_0\| \leq t_k - t_0 = t_k \leq t^*$ and $A(x_k, x_{k-1})$ is invertible by (h_2). We can write by method (7.1.6)

$$x_{k+1} - x_k = -A_k^{-1}(F(x_k) - F(x_{k-1}) - A_{k-1}(x_k - x_{k-1})). \tag{7.2.4}$$

In view of the induction hypothesis (7.2.3), (h_2), (h_3), (h_4), (7.2.1) and (7.2.4), we get in turn that

$$\|x_{k+1} - x_k\| = \left\|A_k^{-1} F(x_k)\right\| = \left\|A_k^{-1} (F(x_k) - F(x_{k-1}) - A_{k-1}(x_k - x_{k-1}))\right\|$$

$$\leq \left\|A_k^{-1}\right\| \|F(x_k) - F(x_{k-1}) - A_{k-1}(x_k - x_{k-1})\| \leq$$

$$l^{-1} l \left(\frac{\alpha_1}{2} \|x_k - x_{k-1}\| + \alpha_2 \|x_{k-1} - x_0\| + \alpha_3\right) \|x_k - x_{k-1}\| \leq \qquad (7.2.5)$$

$$\frac{\alpha_1}{2} (t_k - t_{k-1})^2 + \alpha_2 (t_k - t_{k-1}) t_{k-1} + \alpha_3 (t_k - t_{k-1}) =$$

$$\frac{\alpha_1}{2} (t_k - t_{k-1})^2 + \alpha_2 (t_k - t_{k-1}) t_{k-1} + \alpha_3 (t_k - t_{k-1}) - (t_k - t_{k-1}) + g(t_{k-1}) =$$

$$g(t_k) - (\alpha_1 - \alpha_2)(t_k - t_{k-1}) t_{k-1} \leq$$

$$g(t_k) = t_{k+1} - t_k, \qquad (7.2.6)$$

which completes the induction for estimate (7.2.2).

That is, we have for any k

$$\|x_{k+1} - x_k\| \leq t_{k+1} - t_k \qquad (7.2.7)$$

and

$$\|x_k - x_0\| \leq t_k \leq t^*. \qquad (7.2.8)$$

It follows by (7.2.7) and (7.2.8) that $\{x_k\}$ is a complete sequence in a Banach space B_1 and as such it converges to some $x^* \in \overline{U}(x_0, t^*)$ (since $\overline{U}(x_0, t^*)$ is a closed set). By letting $k \to +\infty$, using (h_1) and (h_2), we get $l^{-1} \lim_{k \to +\infty} \|F(x_k)\| = 0$, so $F(x^*) = 0$.

Let $x^{**} \in \widetilde{U}$ be such that $F(x^{**}) = 0$. We shall show by induction that

$$\left\|x^{**} - x_k\right\| \leq t^* - t_k \text{ for each } k = 0, 1, 2, \dots. \qquad (7.2.9)$$

Estimate (7.2.9) holds for $k = 0$ by the definition of x^{**} and $\widetilde{U}$. Suppose that $\|x^{**} - x_k\| \leq t^* - t_k$. Then, as in (7.2.5), we obtain in turn that

$$\left\|x^{**} - x_{k+1}\right\| = \left\|x^{**} - x_k + A_k^{-1} F(x_k) - A_k^{-1} F(x^{**})\right\| =$$

$$\left\|A_k^{-1} \left(A_k \left(x^{**} - x_k\right) + F(x_k) - F(x^{**})\right)\right\| \leq$$

$$\left\|A_k^{-1}\right\| \left\|F(x^{**}) - F(x_k) - A_k \left(x^{**} - x_k\right)\right\| \leq$$

$$\left(\frac{\alpha_1}{2} \left\|x^{**} - x_k\right\| + \alpha_2 \|x_k - x_0\| + \alpha_3\right) \left\|x^{**} - x_k\right\| \leq$$

$$\left(\frac{\alpha_1}{2}\left(t^* - t_k\right) + \alpha_2 t_k + \alpha_3\right)\left(t^* - t_k\right) =$$

$$\frac{\alpha_1}{2}\left(t^*\right)^2 + \frac{\alpha_1}{2}\left(t_k\right)^2 - \alpha_1 t_k t^* + \alpha_2\left(t^* - t_k\right)t_k + \alpha_3\left(t^* - t_k\right) =$$

$$-\eta + (1 - \alpha_3)t^* + \frac{\alpha_1}{2}t_k^2 - \alpha_1 t_k t^* + \alpha_2 t_k t^* - \alpha_2 t_k^2 + \alpha_3 t^* - \alpha_3 t_k$$

$$= t^* - t_{k+1}, \tag{7.2.10}$$

which completes the induction for estimate (7.2.9). Hence, $\lim_{k \to +\infty} x_k = x^{**}$. But we showed that $\lim_{k \to +\infty} x_k = x^*$, so $x^{**} = x^*$.

Case $\alpha_1 = 0$. Then, we have by (h_3) that $\alpha_2 = 0$ and estimate (7.2.5) gives

$$\|x_{k+1} - x_k\| \le \alpha_3 \|x_k - x_{k-1}\| \le \dots \le \alpha_3^k \|x_1 - x_0\| \le \alpha_3^k \eta \tag{7.2.11}$$

and

$$\|x_{k+1} - x_0\| \le \|x_{k+1} - x_k\| + \|x_k - x_{k-1}\| + \dots + \|x_1 - x_0\|$$

$$\le \frac{1 - \alpha_3^{k+1}}{1 - \alpha_3}\eta < \frac{\eta}{1 - \alpha_3}. \tag{7.2.12}$$

Then, as in the previous case it follows from (7.2.11) and (7.2.12) that

$$\|x_{k+i} - x_k\| \le \frac{1 - \alpha_3^i}{1 - \alpha_3}\alpha_3^k \eta, \tag{7.2.13}$$

so sequence $\{x_k\}$ is complete and x^* solves equation $F(x) = 0$. Finally, the uniqueness part follows from (7.2.10) for $\alpha_1 = \alpha_2 = 0$, since

$$\left\|x^{**} - x_{k+1}\right\| \le \alpha_3 \left\|x^{**} - x_k\right\| \le \alpha_3^{k+1}\left\|x^{**} - x_0\right\| \le \alpha_3^{k+1}\frac{\eta}{1 - \alpha_3}, \tag{7.2.14}$$

which shows again that $\lim_{k \to +\infty} x_k = x^{**}$. ∎

Remark 7.2 (1) Condition (h_2) can be incorporated in (h_3) as follows

(h_3') There exist real numbers $\overline{\alpha}_1, \overline{\alpha}_2, \overline{\alpha}_3$ satisfying $0 \le \overline{\alpha}_2 \le \overline{\alpha}_1$ and $0 \le \overline{\alpha}_3 < 1$ such that for each $x, y \in \Omega$

$$\left\|A(x, y)^{-1}\left[F(x) - F(y) - A(x, y)(x - y)\right]\right\| \le$$

$$\left(\overline{(\alpha_1}/2)\|x - y\| + \overline{\alpha}_2\|y - x_0\| + \overline{\alpha}_3\right)\|x - y\|.$$

Then, (h_3') can replace (h_2) and (h_3) in Theorem 7.1 for $\alpha_1 = \overline{\alpha}_1$, $\overline{\alpha}_2 = \alpha_2$, $\overline{\alpha}_3 = \alpha_3$ and $\Omega_0 = \Omega$. Moreover, notice that $\overline{\alpha}_1 \leq \alpha_1$, $\overline{\alpha}_2 \leq \alpha_1$ and $\overline{\alpha}_3 \leq \alpha_3$, which play a role in the sufficient convergence criterion (h_6), error bounds and the precision of t^* and t^{**}. Condition (h_3) is of Mysowksii-type [11].

(2) Suppose that there exist $l_0 > 0$, $\alpha_4 > 0$ and $L \in \mathcal{L}(B_1, B_2)$ with $L^{-1} \in \mathcal{L}(B_2, B_1)$ such that $\|L^{-1}\| \leq l_0^{-1}$

$$\|A(F)(x, y) - L\| \leq \alpha_4 \text{ for each } x, y \in \Omega$$

and

$$\alpha_5 := l_0^{-1}\alpha_4 < 1.$$

Then, it follows from the Banach lemma on invertible operators [6, 10, 11, 15, 16] and

$$\left\|L^{-1}\right\| \|A(F)(x, y) - L\| \leq l_0^{-1}\alpha_4 = \alpha_5 < 1$$

that $A(F)(x, y)^{-1} \in \mathcal{L}(B_2, B_1)$. Set $l^{-1} = \frac{l_0^{-1}}{1-\alpha_5}$. Then, under these replacements, condition (h_2) is implied, so it can be dropped from the conditions (H).

(3) Clearly method (7.1.5) converges under the conditions (H), since (7.1.6) implies (7.1.5).

(4) Let $R > 0$ and define $R_0 = \sup\{t \in [0, R) : U(x_0, R_0) \subseteq D\}$. Set $\Omega_0 = \overline{U}(x_0, R_0)$. Condition (h_3) can be extended, if the additional term $a_2 \|x - x_0\|$ is inserted inside the parenthesis at the right hand side for some $a_2 \geq 0$. Then, the conclusions of Theorem 7.1 hold in this more general setting, provided that $a_3 = a_2 R_0 + \alpha_3$ replaces α_3 in conditions (h_6) and (h_7).

(5) Concerning the solvability of Eqs. (7.1.6) (or (7.1.5)), we wanted to leave condition (h_4) as uncluttered as possible in conditions (H). We did this because in practice these equations may be solvable in a way other than using the contraction mapping principle already mentioned earlier.

Next, we show the solvability of method (7.1.5) using a stronger version of the contraction mapping principle and based on the conditions (C) :

$(c_1) = (h_1)$.

(c_2) There exist $\gamma_0 \in [0, 1)$, $\gamma_1 \in [0, +\infty)$, $\gamma_2 \in [0, 1)$, $x_0 \in \Omega$ such that for each $x, y, z \in \Omega$

$$\|I + A(x, z) - A(y, z)\| \leq \gamma_0,$$

$$\|A(x, z) - A(y, z)\| \leq \gamma_1\|x - y\|$$

$$\|F(z) + A(x_0, z)(x_0 - z)\| \leq \begin{cases} \gamma_2 \|x_0 - z\| & \text{for } x_0 \neq z \\ \|F(x_0)\| & \text{for } x_0 = z \end{cases}$$

(c_3)

$$\gamma_0 + \gamma_1 \|x_0\| + \gamma_2 \leq 1 \text{ for } \gamma_2 \neq 0,$$

$$\gamma_0 + \gamma_1 \|x_0\| < 1 \text{ for } \gamma_2 = 0,$$

$$\|F(x_0)\| \leq \frac{(1 - (\gamma_0 + \gamma_1 \|x_0\|))^2}{\gamma_1} \text{ for } \gamma_1 \neq 0,$$

$$\gamma_0 < 1 \text{ for } \gamma_1 = 0$$

and

(c_4) $\overline{U}(x_0, r) \subseteq \Omega$, where

$$\frac{\|F(x_0)\|}{1 - (\gamma_0 + \gamma_1 \|x_0\|)} \leq r < \frac{1 - (\gamma_0 + \gamma_1 \|x_0\|)}{\gamma_1} \text{ for } \gamma_1 \neq 0,$$

$$\frac{\|F(x_0)\|}{1 - \gamma_0} \leq r \text{ for } \gamma_1 = 0,$$

$$r < \frac{1 - (\gamma_0 + \gamma_1 \|x_0\|)}{\gamma_1} \text{ for } z = x_0, \gamma_1 \neq 0.$$

Theorem 7.3 *Suppose that the conditions* (C) *are satisfied. Then, for each* $n = 0, 1, 2, \ldots$ *Eq.* (7.1.5) *is unique solvable. Moreover, if* $A_n^{-1} \in \mathcal{L}(B_2, B_1)$, *then Eq.* (7.1.6) *is also uniquely solvable for each* $n = 0, 1, 2, \ldots$

Proof We base the proof on the contraction mapping principle. Let $x, y \in U(x_0, r)$. Then, using (7.1.8) we have in turn by (c_2) that

$$\|P_z(x) - P_z(y)\| = \|(I + A(x, z) - A(y, z))(x - y) - (A(x, z) - A(y, z))z\|$$

$$\leq \|I + A(x, z) - A(y, z)\| \|x - y\| + \|A(x, z) - A(y, z)\| \|z\|$$

$$\leq \gamma_0 \|x - y\| + \gamma_1 (\|z - x_0\| + \|x_0\|) \|x - y\|$$

$$\leq \varphi(\|x - x_0\|) \|x - y\|, \tag{7.2.15}$$

where

$$\varphi(t) = \begin{cases} \gamma_0 + \gamma_1 (t + \|x_0\|) & \text{for } z \neq x_0 \\ \gamma_0 + \gamma_1 \|x_0\| & \text{for } z = x_0. \end{cases} \tag{7.2.16}$$

Notice that $\varphi(t) \in [0, 1)$ for $t \in [0, r]$ by the choice of r in (c_4).

We also have that

$$\|P_z(x) - x_0\| \leq \|P_z(x) - P_z(x_0)\| + \|P_z(x_0) - x_0\|. \tag{7.2.17}$$

If $z = x_0$ in (7.2.17), then we get by (c_3), (c_4) and (7.2.15) that

$$\|P_{x_0}(x) - x_0\| \leq \varphi(\|x - x_0\|) \|x - x_0\| + \|F(x_0)\|$$

$$\leq (\gamma_0 + \gamma_1 \|x_0\|) r + \|F(x_0)\| \leq r. \tag{7.2.18}$$

The existence of $x_1 \in U(x_0, r)$ solving (7.1.5) for $n = 0$ is now established by the contraction mapping principle, (7.2.15) and (7.2.18).

Moreover, if $z \neq x_0$, the last condition in (c_3), (c_3), (c_4) and (7.2.17) give instead of (7.2.18) that

$$\|P_z(x) - x_0\| \leq \varphi(\|x - x_0\|) \|x - x_0\| + \gamma_2 \|x - x_0\|$$

$$\leq (\gamma_0 + \gamma_1 \|x_0\| + \gamma_2) r \leq r. \tag{7.2.19}$$

Then, again by (7.2.15), (7.2.19) and the contraction mapping principle, we guarantee the unique solvability of Eq. (7.1.5) and the existence of a unique sequence $\{x_n\}$ for each $n = 0, 1, 2, \ldots$. Finally, Eq. (7.1.6) is also uniquely solvable by the preceding proof and the condition $A_n^{-1} \in \mathcal{L}(B_2, B_1)$. ∎

Remark 7.4 (a) The gamma conditions can be weakened, if γ_i are replaced by functions $\gamma_i(t)$, $i = 0, 1, 2, 3$. Then, γ_i will appear as $\gamma_i(\|x - x_0\|)$ and $\gamma_i(r)$ in the conditions (C).

(b) Sections 7.2 and 7.3 have an interest independent of Sect. 7.4. However, the results especially of Theorem 7.1 can apply in Abstract Fractional Calculus as suggested in Sect. 7.4.

7.3 Semi-local Convergence for Explicit Methods

Theorem 7.1 is general enough so it can be used to study the semi-local convergence of method (7.1.11), method (7.1.12), method (7.1.13) and method (7.1.14). In particular, for the study of method (7.1.12) (and consequently method (7.1.11)), we use the conditions (H'):

(h_1') $F : \Omega \subset B_1 \rightarrow B_2$ is continuous and $A(F)(x, y) \in \mathcal{L}(B_1, B_2)$ for each $x \in \Omega$.

(h_2') There exist $l > 0$ and $\Omega_0 \subset B_1$ such that $A(F)(x, x)^{-1} \in \mathcal{L}(B_2, B_1)$ and

$$\|A(F)(x, x)^{-1}\| \leq l^{-1}.$$

Set $\Omega_1 = \Omega \cap \Omega_0$.

(h_3') There exist real numbers $\gamma_1, \alpha_2, \gamma_3$ satisfying

$$0 \leq \alpha_2 \leq \gamma_1 \text{ and } 0 \leq \gamma_3$$

such that for each $x, y \in \Omega_1$

$$\|F(x) - F(y) - A(F)(y, y)(x - y)\| \leq$$

$$l\left(\frac{\gamma_1}{2} \|x - y\| + \alpha_2 \|y - x_0\| + \gamma_3\right) \|x - y\|.$$

(h_4') For each $x, y \in \Omega_1$ and some $\gamma_4 \geq 0, \gamma_5 \geq 0$

$$\|A(x, y) - A(y, y)\| \leq l\gamma_4$$

or

$$\|A(x, y) - A(y, y)\| \leq l\gamma_5 \|x - y\|.$$

Set $\alpha_1 = \gamma_1 + \gamma_5$ and $\alpha_3 = \gamma_3 + \gamma_4$, if the second inequation holds or $\alpha_1 = \gamma_1$ and $\alpha_3 = \gamma_3 + \gamma_4$, if the first inequation holds. Further, suppose $0 \leq \alpha_3 < 1$.
    (h_5') There exist $x_0 \in \Omega_0$ and $\eta \geq 0$ such that $A(F)(x_0, x_0)^{-1} \in L(B_2, B_1)$ and

$$\left\|A(F)(x_0, x_0)^{-1} F(x_0)\right\| \leq \eta.$$

$(h_6') = (h_6)$
and
$(h_7') = (h_7)$.
    Then, we can show the following semi-local convergence of method (7.1.12) using the conditions (H') and the preceding notation.

Proposition 7.5 *Suppose that the conditions (H') are satisfied. Then, sequence $\{x_n\}$ generated by method (7.1.12) starting at $x_0 \in \Omega$ is well defined in $U(x_0, t^*)$, remains in $U(x_0, t^*)$ for each $n = 0, 1, 2, \ldots$ and converges to a solution $x^* \in \overline{U}(x_0, t^*)$ of equation $F(x) = 0$. Moreover, if $\alpha_1 \neq 0$, the equation $F(x) = 0$ has a unique solution x^* in $\tilde{U}$, where*

$$\tilde{U} = \begin{cases} \overline{U}(x_0, t^*) \cap \Omega_0, & \text{if } h = \frac{1}{2}(1 - \alpha_3)^2 \\ U(x_0, t^{**}) \cap \Omega_0, & \text{if } h < \frac{1}{2}(1 - \alpha_3)^2 \end{cases}$$

and, if $\alpha_1 = 0$, the solution x^* is unique in $\overline{U}\left(x_0, \frac{\eta}{1-\alpha_3}\right)$, where t^* and t^{**} are given in Theorem 7.1.

Proof Use in the proof of Theorem 7.1 instead of estimate (7.2.5) the analogous estimate

$$\|F(x_k)\| = \|F(x_k) - F(x_{k-1}) - A(x_{k-1}, x_{k-1})(x_k - x_{k-1})\| =$$

$$\left\|[F(x_k) - F(x_{k-1}) - A(x_k, x_{k-1})(x_k - x_{k-1})] +\right.$$

$$(A(x_k, x_{k-1}) - A(x_{k-1}, x_{k-1}))(x_k - x_{k-1})\|$$

$$\le l\left(\frac{\gamma_1}{2}\|x_k - x_{k-1}\| + \alpha_2\|x_{k-1} - x_0\| + \gamma_3\right)\|x_k - x_{k-1}\| +$$

$$\|A(x_k, x_{k-1}) - A(x_{k-1}, x_{k-1})\|\|x_k - x_{k-1}\| \le$$

$$l\left(\frac{\alpha_1}{2}(t_k - t_{k-1})^2 + \alpha_2(t_k - t_{k-1})t_{k-1} + \alpha_3(t_k - t_{k-1})\right),$$

where we used again that $\|x_k - x_{k-1}\| \le t_k - t_{k-1}$, $\|x_{k-1} - x_0\| \le t_{k-1}$ and the condition (h'_4). ∎

Remark 7.6 Comments similar to Remark 7.2 (1)–(3) can follow but for method (7.1.11) and method (7.1.12) instead of method (7.1.5) and method (7.1.6), respectively.

Similarly, for method (7.1.13) and method (7.1.14), we use the conditions (H'') :
$(h''_1) = (h_1)$
$(h''_2) = (h_2)$
(h''_3) There exist real numbers $\alpha_1, \alpha_2, \gamma_3$ satisfying

$$0 \le \alpha_2 \le \alpha_1 \text{ and } 0 \le \gamma_3$$

such that for each $x, y \in \Omega_1$

$$\|F(x) - F(y) - A(F)(x, y)(x - y)\| \le$$

$$l\left(\frac{\alpha_1}{2}\|x - y\| + \alpha_2\|y - x_0\| + \gamma_3\right)\|x - y\|.$$

(h''_4) For each $x, y, z \in \Omega_1$ and some $\gamma_3 \ge 0$

$$\|A(z, y) - A(y, x)\| \le l\delta_3.$$

Set $\alpha_3 = \gamma_3 + \delta_3$ and further suppose $0 \le \alpha_3 < 1$.
(h''_5) There exist $x_{-1} \in \Omega$, $x_0 \in \Omega$ and $\eta \ge 0$ such that $A(F)(x_0, x_{-1})^{-1} \in \mathcal{L}(B_2, B_1)$ and

$$\left\|A(F)(x_0, x_{-1})^{-1}F(x_0)\right\| \le \eta.$$

$(h''_6) = (h_6)$
and
$(h''_7) = (h_7)$.
Then, we can present the following semi-local convergence of method (7.1.14) using the conditions (H'') and the preceding notation.

Proposition 7.7 *Suppose that the conditions (H'') are satisfied. Then, sequence $\{x_n\}$ generated by method (7.1.14) starting at $x_0 \in \Omega$ is well defined in $U(x_0, t^*)$, remains*

in $U(x_0, t^*)$ for each $n = 0, 1, 2, \ldots$ and converges to a solution $x^* \in \overline{U}(x_0, t^*)$
of equation $F(x) = 0$. Moreover, if $\alpha_1 \neq 0$, the equation $F(x) = 0$ has a unique
solution x^* in $\tilde{U}$, where

$$\tilde{U} = \begin{cases} \overline{U}(x_0, t^{**}) \cap \Omega_0, & \text{if } h = \frac{1}{2}(1 - \alpha_3)^2 \\ U(x_0, t^{**}) \cap \Omega_0, & \text{if } h < \frac{1}{2}(1 - \alpha_3)^2 \end{cases}$$

and, if $\alpha_1 = 0$, the solution x^* is unique in $\overline{U}\left(x_0, \frac{\eta}{1-\alpha_3}\right)$, where t^* and t^{**} are given
in Theorem 7.1.

Proof As in Proposition 7.5, use in the proof of Theorem 7.1 instead of estimate
(7.2.5) the analogous estimate

$$\|F(x_k)\| =$$

$$\|F(x_k) - F(x_{k-1}) - A(x_k, x_{k-1})(x_k - x_{k-1})$$

$$+ (A(x_k, x_{k-1}) - A(x_{k-1}, x_{k-2}))(x_k - x_{k-1})\| \le$$

$$\|F(x_k) - F(x_{k-1}) - A(x_k, x_{k-1})(x_k - x_{k-1})\| +$$

$$\|A(x_k, x_{k-1}) - A(x_{k-1}, x_{k-2})\| \, \|x_k - x_{k-1}\|$$

$$\le l\left(\frac{\alpha_1}{2}\|x_k - x_{k-1}\| + \alpha_2\|x_{k-1} - x_0\| + \gamma_3\right)\|x_k - x_{k-1}\| + l\delta_3\|x_k - x_{k-1}\|$$

$$\le l\left(\frac{\alpha_1}{2}(t_k - t_{k-1})^2 + \alpha_2(t_k - t_{k-1})t_{k-1} + \alpha_3(t_k - t_{k-1})\right),$$

where we used again that $\|x_k - x_{k-1}\| \le t_k - t_{k-1}$, $\|x_{k-1} - x_0\| \le t_{k-1}$ and the
condition (h_4'').   ∎

Remark 7.8 Comments similar to Remark 7.2 (1)–(3) can follow but for method
(7.1.13) and method (7.1.14) instead of method (7.1.5) and method (7.1.6), respectively.

In particular, the results of Sect. 7.4 connect to the crucial conditions (h_3') and
(h_4') as follows: According to the definition of method (7.1.12) and using (7.4.1),
(7.4.8) and (7.4.15), we can write

$$\|f(x)\| = \|f(x) - f(y) - A(y)(x - y)\| =$$

$$\left\|[f(x) - f(y) - f'(y)(x - y)] + [f'(y) - A(y)](x - y)\right\| \le$$

$$\left\|f(x) - f(y) - f'(y)(x - y)\right\| + \left\|f'(y) - A(y)\right\| \, \|x - y\| \le$$

$$\frac{\|f''\|_\infty}{2}|x - y|^2 + \frac{1 - \alpha}{2 - \alpha}\|f''\|_\infty \, |y - x_0| \, |x - y|.$$

It follows that $\left(h_3'\right)$ and $\left(h_4'\right)$ hold for

$$\gamma_1 = \frac{\|f''\|_\infty}{l}, \quad \alpha_2 = \frac{1-\alpha}{(2-\alpha)l}\|f''\|_\infty$$

and $\gamma_3 = \gamma_4 = \gamma_5 = 0$. Notice also that $0 \leq \gamma_3$ and $0 \leq \alpha_2 \leq \gamma_1$, since $1-\alpha < 2-\alpha$ and $\gamma_3 = 0$.

7.4 Applications to Abstract Fractional Calculus

Here we deal with Banach space $(X, \|\cdot\|)$ valued functions f of real domain $[c, d]$. All integrals here are of Bochner-type, see [14]. The derivatives of f are defined similarly to numerical ones, see [17], pp. 83–86 and p. 93.

We want to solve numerically $f(x) = 0$.

Let $0 < \alpha < 1$, hence $\lceil \alpha \rceil = 1$, where $\lceil \cdot \rceil$ is the ceiling of the number. Let also $c < a < b < d$, and $f \in C^2([c, d], X)$, with $f'' \neq 0$. Clearly we have (see [12], p. 3)

$$\|f'(x) - f'(y)\| \leq \|f''\|_\infty |x - y|, \ \forall x, y \in [c, d]. \tag{7.4.1}$$

(I) The X-valued left Caputo fractional derivative (see [4]) of f of order $\alpha \in (0, 1)$, anchored at a, is defined as follows:

$$\left(D_{*a}^\alpha f\right)(x) = \frac{1}{\Gamma(1-\alpha)} \int_a^x (x-t)^{-\alpha} f'(t)\, dt, \ \forall x \in [a, d], \tag{7.4.2}$$

while $\left(D_{*a}^\alpha f\right)(x) = 0$, for $c \leq x \leq a$, where Γ is the gamma function.

Next we consider $a < a^* < b$, and $x \in [a^*, b]$, also $x_0 \in (c, a)$.

We define the function

$$A_1(x) := \frac{\Gamma(2-\alpha)}{(x-a)^{1-\alpha}} \left(D_{*a}^\alpha f\right)(x), \ \forall x \in [a^*, b]. \tag{7.4.3}$$

Notice that $A_1(a)$ is undefined.

We see that

$$\left\|A_1(x) - f'(x)\right\| = \left\|\frac{\Gamma(2-\alpha)}{(x-a)^{1-\alpha}} \left(D_{*a}^\alpha f\right)(x) - f'(x)\right\| = \tag{7.4.4}$$

$$\left\|\frac{\Gamma(2-\alpha)}{(x-a)^{1-\alpha}} \frac{1}{\Gamma(1-\alpha)} \int_a^x (x-t)^{-\alpha} f'(t)\, dt - \frac{\Gamma(2-\alpha)}{(x-a)^{1-\alpha}} \frac{(x-a)^{1-\alpha}}{\Gamma(2-\alpha)} f'(x)\right\| =$$

$$\frac{\Gamma(2-\alpha)}{(x-a)^{1-\alpha}} \left\|\frac{1}{\Gamma(1-\alpha)} \int_a^x (x-t)^{-\alpha} f'(t)\, dt - \frac{1}{\Gamma(1-\alpha)} \int_a^x (x-t)^{-\alpha} f'(x)\, dt\right\|$$

(by [1] p. 246, Theorem 11.43)

$$= \frac{(1-\alpha)}{(x-a)^{1-\alpha}} \left\| \int_a^x (x-t)^{-\alpha} \left(f'(t) - f'(x) \right) dt \right\| \tag{7.4.5}$$

(by [8])

$$\leq \frac{(1-\alpha)}{(x-a)^{1-\alpha}} \int_a^x (x-t)^{-\alpha} \left\| f'(t) - f'(x) \right\| dt \overset{(7.4.1)}{\leq}$$

$$\frac{(1-\alpha)\left\| f'' \right\|_\infty}{(x-a)^{1-\alpha}} \int_a^x (x-t)^{-\alpha} (x-t) \, dt = \frac{(1-\alpha)\left\| f'' \right\|_\infty}{(x-a)^{1-\alpha}} \int_a^x (x-t)^{1-\alpha} \, dt =$$

$$\frac{(1-\alpha)\left\| f'' \right\|_\infty}{(x-a)^{1-\alpha}} \frac{(x-a)^{2-\alpha}}{2-\alpha} = \frac{(1-\alpha)}{(2-\alpha)} \left\| f'' \right\|_\infty (x-a). \tag{7.4.6}$$

We have proved that

$$\left\| A_1(x) - f'(x) \right\| \leq \left(\frac{1-\alpha}{2-\alpha} \right) \left\| f'' \right\|_\infty (x-a) \leq \left(\frac{1-\alpha}{2-\alpha} \right) \left\| f'' \right\|_\infty (b-a),$$
$$\tag{7.4.7}$$

$\forall x \in [a^*, b]$.

In particular, it holds that

$$\left\| A_1(x) - f'(x) \right\| \leq \left(\frac{1-\alpha}{2-\alpha} \right) \left\| f'' \right\|_\infty (x-x_0), \tag{7.4.8}$$

where $x_0 \in (c, a)$, $\forall x \in [a^*, b]$.

(II) The X-valued right Caputo fractional derivative (see [3]) of f of order $\alpha \in (0, 1)$, anchored at b, is defined as follows:

$$\left(D_{b-}^\alpha f \right)(x) = \frac{-1}{\Gamma(1-\alpha)} \int_x^b (t-x)^{-\alpha} f'(t) \, dt, \quad \forall x \in [c, b], \tag{7.4.9}$$

while $\left(D_{b-}^\alpha f \right)(x) = 0$, for $d \geq x \geq b$.

Next consider $a < b^* < b$, and $x \in [a, b^*]$, also $x_0 \in (b, d)$.

We define the function

$$A_2(x) := -\frac{\Gamma(2-\alpha)}{(b-x)^{1-\alpha}} \left(D_{b-}^\alpha f \right)(x), \quad \forall x \in [a, b^*]. \tag{7.4.10}$$

Notice that $A_2(b)$ is undefined.

We see that

$$\left\| A_2(x) - f'(x) \right\| = \left\| -\frac{\Gamma(2-\alpha)}{(b-x)^{1-\alpha}} \left(D_{b-}^\alpha f \right)(x) - f'(x) \right\| =$$

$$\left\| \frac{\Gamma(2-\alpha)}{(b-x)^{1-\alpha}} \frac{1}{\Gamma(1-\alpha)} \int_x^b (t-x)^{-\alpha} f'(t)\,dt - f'(x) \right\| = \qquad (7.4.11)$$

$$\left\| \frac{\Gamma(2-\alpha)}{(b-x)^{1-\alpha}} \frac{1}{\Gamma(1-\alpha)} \int_x^b (t-x)^{-\alpha} f'(t)\,dt - \frac{\Gamma(2-\alpha)}{(b-x)^{1-\alpha}} \frac{(b-x)^{1-\alpha}}{\Gamma(2-\alpha)} f'(x) \right\| =$$

$$\frac{\Gamma(2-\alpha)}{(b-x)^{1-\alpha}} \left\| \frac{1}{\Gamma(1-\alpha)} \int_x^b (t-x)^{-\alpha} f'(t)\,dt - \frac{1}{\Gamma(1-\alpha)} \int_x^b (t-x)^{-\alpha} f'(x)\,dt \right\| =$$
$$\qquad (7.4.12)$$

$$\frac{\Gamma(2-\alpha)}{(b-x)^{1-\alpha}} \frac{1}{\Gamma(1-\alpha)} \left\| \int_x^b (t-x)^{-\alpha} \left(f'(t) - f'(x) \right) dt \right\| \le$$

$$\frac{(1-\alpha)}{(b-x)^{1-\alpha}} \int_x^b (t-x)^{-\alpha} \left\| f'(t) - f'(x) \right\| dt \overset{(7.4.1)}{\le}$$

$$\frac{(1-\alpha)\left\| f'' \right\|_\infty}{(b-x)^{1-\alpha}} \int_x^b (t-x)^{-\alpha} (t-x)\,dt = \frac{(1-\alpha)\left\| f'' \right\|_\infty}{(b-x)^{1-\alpha}} \int_x^b (t-x)^{1-\alpha}\,dt =$$
$$\qquad (7.4.13)$$

$$\frac{(1-\alpha)\left\| f'' \right\|_\infty}{(b-x)^{1-\alpha}} \frac{(b-x)^{2-\alpha}}{2-\alpha} = \frac{(1-\alpha)\left\| f'' \right\|_\infty}{(2-\alpha)} (b-x).$$

We have proved that

$$\left\| A_2(x) - f'(x) \right\| \le \left(\frac{1-\alpha}{2-\alpha} \right) \left\| f'' \right\|_\infty (b-x) \le \left(\frac{1-\alpha}{2-\alpha} \right) \left\| f'' \right\|_\infty (b-a),$$
$$\qquad (7.4.14)$$

$\forall\, x \in [a, b^*]$.

In particular, it holds that

$$\left\| A_2(x) - f'(x) \right\| \le \left(\frac{1-\alpha}{2-\alpha} \right) \left\| f'' \right\|_\infty (x_0 - x), \qquad (7.4.15)$$

where $x_0 \in (b, d)$, $\forall\, x \in [a, b^*]$.

References

1. C.D. Aliprantis, K.C. Border, *Infinite Dimensional Analysis* (Springer, New York, 2006)
2. S. Amat, S. Busquier, S. Plaza, Chaotic dynamics of a third-order Newton-type method. J. Math. Anal. Appl. **366**(1), 164–174 (2010)
3. G.A. Anastassiou, Strong right fractional calculus for banach space valued functions. Rev. Proyecc. **36**(1), 149–186 (2017)

4. G.A. Anastassiou, A strong fractional calculus theory for Banach space valued functions. Nonlinear Funct. Anal. Appl. (Korea), (accepted for publication, 2017)
5. G.A. Anastassiou, I.K. Argyros, *Generalized Iterative procedures and their applications to Banach space valued functions in abstract fractional calculus* (submitted, 2017)
6. I.K. Argyros, A unifying local-semilocal convergence analysis and applications for two-point Newton-like methods in Banach space. J. Math. Anal. Appl. **298**, 374–397 (2004)
7. I.K. Argyros, A. Magréñan, *Iterative Methods and their Dynamics with Applications* (CRC Press, New York, 2017)
8. Bochner integral, Encyclopedia of Mathematics, http://www.encyclopediaofmath.org/index.php?title=Bochner_integral&oldid=38659
9. M. Edelstein, On fixed and periodic points under contractive mappings. J. London Math. Soc. **37**, 74–79 (1962)
10. J.A. Ezquerro, J.M. Gutierrez, M.A. Hernandez, N. Romero, M.J. Rubio, The Newton method: from Newton to Kantorovich (Spanish). Gac. R. Soc. Mat. Esp. **13**, 53–76 (2010)
11. L.V. Kantorovich, G.P. Akilov, *Functional Analysis in Normed Spaces* (Pergamon Press, New York, 1982)
12. G.E. Ladas, V. Lakshmikantham, *Differential Equations in Abstract Spaces* (Academic Press, New York, 1972)
13. A. Magréñan, A new tool to study real dynamics: the convergence plane. Appl. Math. Comput. **248**, 215–224 (2014)
14. J. Mikusinski, *The Bochner Integral* (Academic Press, New York, 1978)
15. F.A. Potra, V. Pták, *Nondiscrete Induction and Iterative Processes* (Pitman Publication, London, 1984)
16. P.D. Proinov, New general convergence theory for iterative processes and its applications to Newton-Kantorovich type theorems. J. Complex. **26**, 3–42 (2010)
17. G.E. Shilov, *Elementary Functional Analysis* (Dover Publications Inc., New York, 1996)

Chapter 8
Approximate Solutions of Equations in Abstract g-Fractional Calculus

The novelty of this chapter is the design of suitable iterative methods for generating a sequence approximating solutions of equations on Banach spaces. Applications of the semi-local convergence are suggested including Banach space valued functions of fractional calculus, where all integrals are of Bochner-type. It follows [6].

8.1 Introduction

Sections 8.1–8.2 are prerequisites for Sect. 8.3.

Let B_1, B_2 stand for Banach spaces and let Ω stand for an open subset of B_1. Let also $U(z, \xi) := \{u \in B_1 : \|u - z\| < \xi\}$ and let $\overline{U}(z, \xi)$ stand for the closure of $U(z, \xi)$.

Many problems in Computational Sciences, Engineering, Mathematical Chemistry, Mathematical Physics, Mathematical Economics and other disciplines can written like

$$F(x) = 0 \tag{8.1.1}$$

using Mathematical Modeling [1–18], where $F : \Omega \to B_2$ is a continuous operator. The solution x^* of Eq. (8.1.1) is needed in closed form. However, this is achieved only in special cases. That explains why most solution methods for such equations are usually iterative. There is a plethora of iterative methods for solving Eq. (8.1.1), more the [2, 7, 8, 10–14, 16, 17].

Newton's method [7, 8, 12, 16, 17]:

$$x_{n+1} = x_n - F'(x_n)^{-1} F(x_n). \tag{8.1.2}$$

Secant method:

$$x_{n+1} = x_n - [x_{n-1}, x_n; F]^{-1} F(x_n), \tag{8.1.3}$$

© Springer International Publishing AG 2018
G. A. Anastassiou and I. K. Argyros, *Functional Numerical Methods: Applications to Abstract Fractional Calculus*, Studies in Systems, Decision and Control 130, https://doi.org/10.1007/978-3-319-69526-6_8

where $[\cdot, \cdot; F]$ denotes a divided difference of order one on $\Omega \times \Omega$ [8, 16, 17].

Newton-like method:

$$x_{n+1} = x_n - E_n^{-1} F(x_n), \qquad (8.1.4)$$

where $E_n = E(F)(x_n)$ and $E : \Omega \to \mathcal{L}(B_1, B_2)$ the space of bounded linear operators from B_1 into B_2. Other methods can be found in [8, 12, 16, 17] and the references therein.

In the present study we consider the new method defined for each $n = 0, 1, 2, \ldots$ by

$$x_{n+1} = G(x_n)$$

$$G(x_{n+1}) = G(x_n) - A_n^{-1} F(x_n), \qquad (8.1.5)$$

where $x_0 \in \Omega$ is an initial point, $G : B_3 \to \Omega$ (B_3 a Banach space), $A_n = A(F)(x_{n+1}, x_n) = A(x_{n+1}, x_n)$ and $A : \Omega \times \Omega \to \mathcal{L}(B_1, B_2)$. Method (8.1.5) generates a sequence which we shall show converges to x^* under some Lipschitz-type conditions (to be precised in Sect. 8.2). Although method (8.1.5) (and Sect. 8.2) is of independent interest, it is nevertheless designed especially to be used in g-Abstract Fractional Calculus (to be precised in Sect. 8.3). As far as we know such iterative methods have not yet appeared in connection to solve equations in Abstract Fractional Calculus.

In this chapter we present the semi-local convergence of method (8.1.5) in Sect. 8.2. Some applications to Abstract g-Fractional Calculus are suggested in Sect. 8.3 on a certain Banach space valued functions, where all the integrals are of Bochner-type [9, 15].

8.2 Semi-local Convergence Analysis

We present the semi-local convergence analysis of method (8.1.5) using conditions (M):

(m_1) $F : \Omega \subset B_1 \to B_2$ is continuous, $G : B_3 \to \Omega$ is continuous and $A(x, y) \in \mathcal{L}(B_1, B_2)$ for each $(x, y) \in \Omega \times \Omega$.

(m_2) There exist $\beta > 0$ and $\Omega_0 \subset B_1$ such that $A(x, y)^{-1} \in \mathcal{L}(B_2, B_1)$ for each $(x, y) \in \Omega_0 \times \Omega_0$ and

$$\left\| A(x, y)^{-1} \right\| \leq \beta^{-1}.$$

Set $\Omega_1 = \Omega \cap \Omega_0$.

(m_3) There exists a continuous and nondecreasing function $\psi : [0, +\infty)^3 \to [0, +\infty)$ such that for each $x, y \in \Omega_1$

$$\| F(x) - F(y) - A(x, y)(G(x) - G(y)) \| \leq$$

$$\beta \psi(\|x - y\|, \|x - x_0\|, \|y - x_0\|) \|G(x) - G(y)\|.$$

(m_4) There exists a continuous and nondecreasing function $\psi_0 : [0, +\infty) \to [0, +\infty)$ such that for each $x \in \Omega_1$

$$\|G(x) - G(x_0)\| \le \psi_0(\|x - x_0\|)\|x - x_0\|.$$

(m_5) For $x_0 \in \Omega_0$ and $x_1 = G(x_0) \in \Omega_0$ there exists $\eta \ge 0$ such that

$$\left\|A(x_1, x_0)^{-1} F(x_0)\right\| \le \eta.$$

(m_6) There exists $s > 0$ such that

$$\psi(\eta, s, s) < 1,$$

$$\psi_0(s) < 1$$

and

$$\|G(x_0) - x_0\| \le s \le \frac{\eta}{1 - q_0},$$

where $q_0 = \psi(\eta, s, s)$.

(m_7) $\overline{U}(x_0, s) \subset \Omega$.

Next, we present the semi-local convergence analysis for method (8.1.5) using the conditions (M) and the preceding notation.

Theorem 8.1 *Assume that the conditions (M) hold. Then, sequence $\{x_n\}$ generated by method (8.1.5) starting at $x_0 \in \Omega$ is well defined in $U(x_0, s)$, remains in $U(x_0, s)$ for each $n = 0, 1, 2, \ldots$ and converges to a solution $x^* \in \overline{U}(x_0, s)$ of equation $F(x) = 0$. The limit point x^* is the unique solution of equation $F(x) = 0$ in $\overline{U}(x_0, s)$.*

Proof By the definition of s and (m_5), we have $x_1 \in U(x_0, s)$. The proof is based on mathematical induction on k. Suppose that $\|x_k - x_{k-1}\| \le q_0^{k-1}\eta$ and $\|x_k - x_0\| \le s$.

We get by (8.1.5), $(m_2) - (m_5)$ in turn that

$$\|G(x_{k+1}) - G(x_k)\| = \left\|A_k^{-1} F(x_k)\right\| =$$

$$\left\|A_k^{-1}(F(x_k) - F(x_{k-1}) - A_{k-1}(G(x_k) - G(x_{k-1})))\right\|$$

$$\le \left\|A_k^{-1}\right\| \|F(x_k) - F(x_{k-1}) - A_{k-1}(G(x_k) - G(x_{k-1}))\| \le$$

$$\beta^{-1}\beta\psi(\|x_k - x_{k-1}\|, \|x_{k-1} - x_0\|, \|y_k - x_0\|)\|G(x_k) - G(x_{k-1})\| \le$$

$$\psi(\eta, s, s)\|G(x_k) - G(x_{k-1})\| = q_0\|G(x_k) - G(x_{k-1})\| \le q_0^k\|x_1 - x_0\| \le \tag{8.2.1}$$

$$q_0^k\eta$$

and by (m_6)

$$\|x_{k+1} - x_0\| = \|G(x_k) - x_0\| \le \|G(x_k) - G(x_0)\| + \|G(x_0) - x_0\|$$

$$\le \psi_0(\|x_k - x_0\|)\|x_k - x_0\| + \|G(x_0) - x_0\|$$

$$\le \psi_0(s)s + \|G(x_0) - x_0\| \le s.$$

The induction is completed. Moreover, we have by (8.2.1) that for $m = 0, 1, 2, \ldots$

$$\|x_{k+m} - x_k\| \le \frac{1 - q_0^m}{1 - q_0} q_0^k \eta.$$

It follows from the preceding inequation that sequence $\{G(x_k)\}$ is complete in a Banach space B_1 and as such it converges to some $x^* \in \overline{U}(x_0, s)$ (since $\overline{U}(x_0, s)$ is a closed ball). By letting $k \to +\infty$ in (8.2.1) we get $F(x^*) = 0$. We also get by (8.1.5) that $G(x^*) = x^*$. To show the uniqueness part, let $x^{**} \in U(x_0, s)$ be a solution of equation $F(x) = 0$ and $G(x^{**}) = x^{**}$. By using (8.1.5), we obtain in turn that

$$\left\|x^{**} - G(x_{k+1})\right\| = \left\|x^{**} - G(x_k) + A_k^{-1} F(x_k) - A_k^{-1} F(x^{**})\right\| \le$$

$$\left\|A_k^{-1}\right\| \left\|F(x^{**}) - F(x_k) - A_k(G(x^{**}) - G(x_k))\right\| \le$$

$$\beta^{-1} \beta \psi_0\left(\left\|x^{**} - x_k\right\|, \|x_{k+1} - x_0\|, \|x_k - x_0\|\right) \left\|G(x^{**}) - G(x_k)\right\| \le$$

$$q_0 \left\|G(x^{**}) - G(x_k)\right\| \le q_0^{k+1} \left\|x^{**} - x_0\right\|,$$

so $\lim_{k \to +\infty} x_k = x^{**}$. We have shown that $\lim_{k \to +\infty} x_k = x^*$, so $x^* = x^{**}$. ∎

Remark 8.2 (1) Condition (m_2) can become part of condition (m_3) by considering
$(m_3)'$ There exists a continuous and nondecreasing function $\varphi : [0, +\infty)^3 \to [0, +\infty)$ such that for each $x, y \in \Omega_1$

$$\left\|A(x, y)^{-1} [F(x) - F(y) - A(x, y)(G(x) - G(y))]\right\| \le$$

$$\varphi(\|x - y\|, \|x - x_0\|, \|y - x_0\|) \|G(x) - G(y)\|.$$

Notice that

$$\varphi(u_1, u_2, u_3) \le \psi(u_1, u_2, u_3)$$

for each $u_1 \ge 0$, $u_2 \ge 0$ and $u_3 \ge 0$. Similarly, a function φ_1 can replace ψ_1 for the uniqueness of the solution part. These replacements are of Mysovskii-type [7, 12, 16] and influence the weaking of the convergence criterion in (m_6), error bounds and the precision of s.

(2) Suppose that there exist $\beta > 0$, $\beta_1 > 0$ and $L \in \mathcal{L}(B_1, B_2)$ with $L^{-1} \in \mathcal{L}(B_2, B_1)$ such that

$$\|L^{-1}\| \leq \beta^{-1}$$

$$\|A(x, y) - L\| \leq \beta_1$$

and

$$\beta_2 := \beta^{-1}\beta_1 < 1.$$

Then, it follows from the Banach lemma on invertible operators [12], and

$$\|L^{-1}\| \|A(x, y) - L\| \leq \beta^{-1}\beta_1 = \beta_2 < 1$$

that $A(x, y)^{-1} \in \mathcal{L}(B_2, B_1)$. Let $\beta = \frac{\beta^{-1}}{1 - \beta_2}$. Then, under these replacements, condition (m_2) is implied, therefore it can be dropped from the conditions (M).

Remark 8.3 Section 8.2 has an interest independent of Sect. 8.3. It is worth noticing that the results especially of Theorem 8.1 can apply in Abstract g-Fractional Calculus as illustrated in Sect. 8.3. By specializing function ψ, we can apply the results of say Theorem 8.1 in the examples suggested in Sect. 8.3. In particular for (8.3.33), we choose for $u_1 \geq 0$, $u_2 \geq 0$, $u_3 \geq 0$

$$\psi(u_1, u_2, u_3) = \frac{\lambda\mu_1^\alpha}{\beta\Gamma(\alpha)(\alpha + 1)},$$

if $|g(x) - g(y)| \leq \mu_1$ for each $x, y \in [a, b]$;

$$\psi(u_1, u_2, u_3) = \frac{\lambda\mu_2^\alpha}{\beta\Gamma(\alpha)(\alpha + 1)},$$

if $|g(x) - g(y)| \leq \xi_2 \|x - y\|$ for each $x, y \in [a, b]$ and $\mu_2 = \xi_2 |b - a|$;

$$\psi(u_1, u_2, u_3) = \frac{\lambda\mu_3^\alpha}{\beta\Gamma(\alpha)(\alpha + 1)},$$

if $|g(x)| \leq \xi_3$ for each $x, y \in [a, b]$ and $\mu_3 = 2\xi_3$, where α, λ and Γ are given in Sect. 8.3.

Other choices of function ψ are also possible.

Notice that with these choices of function ψ and $f = F$ and $g = G$, crucial condition (m_3) is satisfied, which justifies our definition of method (8.1.5). We can provide similar choices for the other examples of Sect. 8.3.

8.3   Applications to X-valued g-Fractional Calculus

Here we deal with Banach space $(X, \|\cdot\|)$ valued functions f of real domain $[a, b]$. All integrals here are of Bochner-type, see [15]. The derivatives of f are defined similarly to numerical ones, see [18], pp. 83–86 and p. 93.

Here both backgrounds needed come from [5].

(I) We need

Definition 8.4 ([5]) Let $\alpha > 0$, $\lceil \alpha \rceil = n$, $\lceil \cdot \rceil$ the ceiling of the number. Let $f \in C^n([a, b], X)$, where $[a, b] \subset \mathbb{R}$, and $(X, \|\cdot\|)$ is a Banach space. Let $g \in C^1([a, b])$, strictly increasing, such that $g^{-1} \in C^n([g(a), g(b)])$.

We define the left generalized g-fractional derivative X-valued of f of order α as follows:

$$\left(D_{a+;g}^{\alpha} f\right)(x) := \frac{1}{\Gamma(n-\alpha)} \int_a^x (g(x) - g(t))^{n-\alpha-1} g'(t) \left(f \circ g^{-1}\right)^{(n)} (g(t)) \, dt,$$

$(8.3.1)$

$\forall\, x \in [a, b]$. The last integral is of Bochner type.

If $\alpha \notin \mathbb{N}$, by [5], we have that $\left(D_{a+;g}^{\alpha} f\right) \in C([a, b], X)$.

We see that

$$\left(J_{a;g}^{n-\alpha}\left(\left(f \circ g^{-1}\right)^{(n)} \circ g\right)\right)(x) = \left(D_{a+;g}^{\alpha} f\right)(x), \forall x \in [a, b].$$

$(8.3.2)$

We set

$$D_{a+;g}^{n} f(x) := \left(\left(f \circ g^{-1}\right)^{n} \circ g\right)(x) \in C([a, b], X), n \in \mathbb{N},$$

$(8.3.3)$

$$D_{a+;g}^{0} f(x) = f(x), \quad \forall\, x \in [a, b].$$

When $g = id$, then

$$D_{a+;g}^{\alpha} f = D_{a+;id}^{\alpha} f = D_{*a}^{\alpha} f,$$

the usual left X-valued Caputo fractional derivative, see [4].

We need the X-valued left general fractional Taylor's formula.

Theorem 8.5 ([5]) Let $\alpha > 0$, $n = \lceil \alpha \rceil$, and $f \in C^n([a, b], X)$, where $[a, b] \subset \mathbb{R}$ and $(X, \|\cdot\|)$ is a Banach space. Let $g \in C^1([a, b])$, strictly increasing, such that $g^{-1} \in C^n([g(a), g(b)])$, $a \leq x \leq b$. Then

$$f(x) = f(a) + \sum_{i=1}^{n-1} \frac{(g(x) - g(a))^i}{i!} \left(f \circ g^{-1}\right)^{(i)} (g(a)) +$$

$$\frac{1}{\Gamma(\alpha)} \int_a^x (g(x) - g(t))^{\alpha-1} g'(t) \left(D_{a+;g}^{\alpha} f\right)(t) \, dt =$$

$$f(a) + \sum_{i=1}^{n-1} \frac{(g(x) - g(a))^i}{i!} \left(f \circ g^{-1} \right)^{(i)} (g(a)) + \tag{8.3.4}$$

$$\frac{1}{\Gamma(\alpha)} \int_{g(a)}^{g(x)} (g(x) - z)^{\alpha-1} \left(\left(D_{a+;g}^\alpha f \right) \circ g^{-1} \right) (z) \, dz.$$

The remainder of (8.3.4) is a continuous function in $x \in [a, b]$.

Here we are going to operate more generally. We consider $f \in C^n([a, b], X)$. We define the following X-valued left g-fractional derivative of f of order α as follows:

$$\left(D_{y+;g}^\alpha f \right)(x) := \frac{1}{\Gamma(n-\alpha)} \int_y^x (g(x) - g(t))^{n-\alpha-1} g'(t) \left(f \circ g^{-1} \right)^{(n)} (g(t)) \, dt,$$
$$\tag{8.3.5}$$

for any $a \le y \le x \le b$;

$$D_{y+;g}^n f(x) = \left(\left(f \circ g^{-1} \right)^{(n)} \circ g \right) (x), \forall \, x, y \in [a, b], \tag{8.3.6}$$

and

$$D_{y+;g}^0 f(x) = f(x), \forall \, x \in [a, b]. \tag{8.3.7}$$

For $\alpha > 0$, $\alpha \notin \mathbb{N}$, by convention we set that

$$\left(D_{y+;g}^\alpha f \right)(x) = 0, \text{ for } x < y, \ \forall \, x, y \in [a, b]. \tag{8.3.8}$$

Similarly, we define

$$\left(D_{x+;g}^\alpha f \right)(y) := \frac{1}{\Gamma(n-\alpha)} \int_x^y (g(y) - g(t))^{n-\alpha-1} g'(t) \left(f \circ g^{-1} \right)^{(n)} (g(t)) \, dt,$$
$$\tag{8.3.9}$$

for any $a \le x \le y \le b$;

$$D_{x+;g}^n f(y) = \left(\left(f \circ g^{-1} \right)^{(n)} \circ g \right) (y), \forall \, x, y \in [a, b], \tag{8.3.10}$$

and

$$D_{x+;g}^0 f(y) = f(y), \forall \, y \in [a, b]. \tag{8.3.11}$$

For $\alpha > 0$, $\alpha \notin \mathbb{N}$, by convention we set that

$$\left(D_{x+;g}^\alpha f \right)(y) = 0, \text{ for } y < x, \forall \, x, y \in [a, b]. \tag{8.3.12}$$

We get that (see [9])

$$\left\| \left(D_{a+;g}^{\alpha} f \right)(x) \right\| \leq \frac{1}{\Gamma(n-\alpha)} \int_{a}^{x} (g(x) - g(t))^{n-\alpha-1} \, g'(t) \left\| \left(f \circ g^{-1} \right)^{(n)} (g(t)) \right\| dt \tag{8.3.13}$$

$$\leq \frac{\left\| \left(f \circ g^{-1} \right)^{(n)} \circ g \right\|_{\infty,[a,b]}}{\Gamma(n-\alpha)} \int_{a}^{x} (g(x) - g(t))^{n-\alpha-1} \, g'(t) \, dt =$$

$$\frac{\left\| \left(f \circ g^{-1} \right)^{(n)} \circ g \right\|_{\infty,[a,b]}}{\Gamma(n-\alpha+1)} (g(x) - g(a))^{n-\alpha} \leq$$

$$\frac{\left\| \left(f \circ g^{-1} \right)^{(n)} \circ g \right\|_{\infty,[a,b]}}{\Gamma(n-\alpha+1)} (g(b) - g(a))^{n-\alpha}, \forall \, x \in [a,b]. \tag{8.3.14}$$

That is

$$\left(D_{a+;g}^{\alpha} f \right)(a) = 0, \tag{8.3.15}$$

and

$$\left(D_{y+;g}^{\alpha} f \right)(y) = \left(D_{x+;g}^{\alpha} f \right)(x) = 0, \forall \, x, y \in [a,b]. \tag{8.3.16}$$

Thus when $\alpha > 0$, $\alpha \notin \mathbb{N}$, both $D_{y+;g}^{\alpha} f$, $D_{x+;g}^{\alpha} f \in C([a,b], X)$, (see [5]). Hence by Theorem 8.5 we obtain

$$f(x) - f(y) = \sum_{k=1}^{n-1} \frac{\left(f \circ g^{-1} \right)^{(k)} (g(y))}{k!} (g(x) - g(y))^{k} +$$

$$\frac{1}{\Gamma(\alpha)} \int_{y}^{x} (g(x) - g(t))^{\alpha-1} \, g'(t) \left(D_{y+;g}^{\alpha} f \right)(t) \, dt, \forall \, x \in [y,b], \tag{8.3.17}$$

and

$$f(y) - f(x) = \sum_{k=1}^{n-1} \frac{\left(f \circ g^{-1} \right)^{(k)} (g(x))}{k!} (g(y) - g(x))^{k} +$$

$$\frac{1}{\Gamma(\alpha)} \int_{x}^{y} (g(y) - g(t))^{\alpha-1} \, g'(t) \left(D_{x+;g}^{\alpha} f \right)(t) \, dt, \forall \, y \in [x,b], \tag{8.3.18}$$

We define also the following X-valued linear operator

$$(A_1(f))(x, y) :=$$

$$
\begin{cases}
\sum_{k=1}^{n-1} \frac{(f \circ g^{-1})^{(k)}(g(y))}{k!} (g(x) - g(y))^{k-1} + \left(D_{y+;g}^\alpha f\right)(x) \frac{(g(x)-g(y))^{\alpha-1}}{\Gamma(\alpha+1)}, & \text{for } x > y, \\
\sum_{k=1}^{n-1} \frac{(f \circ g^{-1})^{(k)}(g(x))}{k!} (g(y) - g(x))^{k-1} + \left(D_{x+;g}^\alpha f\right)(y) \frac{(g(y)-g(x))^{\alpha-1}}{\Gamma(\alpha+1)}, & \text{for } x < y, \\
f^{(n)}(x), & \text{when } x = y,
\end{cases}
$$

(8.3.19)

$\forall \, x, y \in [a, b]; \alpha > 0, n = \lceil \alpha \rceil$.

We may assume that (see [13], p. 3)

$$\|(A_1(f))(x, x) - (A_1(f))(y, y)\| = \left\| f^{(n)}(x) - f^{(n)}(y) \right\| \qquad (8.3.20)$$

$$\left\| \left(f^{(n)} \circ g^{-1}\right)(g(x)) - \left(f^{(n)} \circ g^{-1}\right)(g(y)) \right\| \le \Phi \left| g(x) - g(y) \right|, \forall \, x, y \in [a, b];$$

where $\Phi > 0$.

We estimate and have

(i) case of $x > y$:

$$\| f(x) - f(y) - (A_1(f))(x, y)(g(x) - g(y)) \| =$$

$$\left\| \frac{1}{\Gamma(\alpha)} \int_y^x (g(x) - g(t))^{\alpha-1} g'(t) \left(D_{y+;g}^\alpha f\right)(t) \, dt - \right.$$

$$\left. \left(D_{y+;g}^\alpha f\right)(x) \frac{(g(x) - g(y))^\alpha}{\Gamma(\alpha+1)} \right\| \qquad (8.3.21)$$

(by [1] p. 426, Theorem 11.43)

$$= \frac{1}{\Gamma(\alpha)} \left\| \int_y^x (g(x) - g(t))^{\alpha-1} g'(t) \left(\left(D_{y+;g}^\alpha f\right)(t) - \left(D_{y+;g}^\alpha f\right)(x)\right) dt \right\|$$

(by [9])

$$\le \frac{1}{\Gamma(\alpha)} \int_y^x (g(x) - g(t))^{\alpha-1} g'(t) \left\| \left(D_{y+;g}^\alpha f\right)(t) - \left(D_{y+;g}^\alpha f\right)(x) \right\| dt \quad (8.3.22)$$

(we assume that

$$\left\| \left(D_{y+;g}^\alpha f\right)(t) - \left(D_{y+;g}^\alpha f\right)(x) \right\| \le \lambda_1 \left| g(t) - g(x) \right|, \qquad (8.3.23)$$

$\forall \, t, x, y \in [a, b] : x \ge t \ge y; \lambda_1 > 0)$

$$\leq \frac{\lambda_1}{\Gamma(\alpha)} \int_y^x (g(x) - g(t))^{\alpha-1} g'(t) (g(x) - g(t)) \, dt =$$

$$\frac{\lambda_1}{\Gamma(\alpha)} \int_y^x (g(x) - g(t))^{\alpha} g'(t) \, dt = \frac{\lambda_1}{\Gamma(\alpha)} \frac{(g(x) - g(y))^{\alpha+1}}{(\alpha+1)}. \qquad (8.3.24)$$

We have proved that

$$\|f(x) - f(y) - (A_1(f))(x, y)(g(x) - g(y))\| \leq$$

$$\frac{\lambda_1}{\Gamma(\alpha)} \frac{(g(x) - g(y))^{\alpha+1}}{(\alpha+1)}, \qquad (8.3.25)$$

$\forall \, x, y \in [a, b] : x > y$.

(ii) case of $y > x$: We have that

$$\|f(x) - f(y) - (A_1(f))(x, y)(g(x) - g(y))\| = \qquad (8.3.26)$$

$$\|f(y) - f(x) - (A_1(f))(x, y)(g(y) - g(x))\| =$$

$$\left\| \frac{1}{\Gamma(\alpha)} \int_x^y (g(y) - g(t))^{\alpha-1} g'(t) \left(D_{x+;g}^{\alpha} f\right)(t) \, dt - \right.$$

$$\left. \left(D_{x+;g}^{\alpha} f\right)(y) \frac{(g(y) - g(x))^{\alpha}}{\Gamma(\alpha+1)} \right\| =$$

$$\frac{1}{\Gamma(\alpha)} \left\| \int_x^y (g(y) - g(t))^{\alpha-1} g'(t) \left(\left(D_{x+;g}^{\alpha} f\right)(t) - \left(D_{x+;g}^{\alpha} f\right)(y)\right) dt \right\| \leq$$

$$\hspace{10cm} (8.3.27)$$

$$\frac{1}{\Gamma(\alpha)} \int_x^y (g(y) - g(t))^{\alpha-1} g'(t) \left\| \left(D_{x+;g}^{\alpha} f\right)(t) - \left(D_{x+;g}^{\alpha} f\right)(y) \right\| dt \qquad (8.3.28)$$

(we assume here that

$$\left\| \left(D_{x+;g}^{\alpha} f\right)(t) - \left(D_{x+;g}^{\alpha} f\right)(y) \right\| \leq \lambda_2 |g(t) - g(y)|, \qquad (8.3.29)$$

$\forall \, t, y, x \in [a, b] : y \geq t \geq x; \lambda_2 > 0$)

$$\leq \frac{\lambda_2}{\Gamma(\alpha)} \int_x^y (g(y) - g(t))^{\alpha-1} g'(t) (g(y) - g(t)) \, dt = \qquad (8.3.30)$$

$$\frac{\lambda_2}{\Gamma(\alpha)} \int_x^y (g(y) - g(t))^{\alpha} g'(t) \, dt = \frac{\lambda_2}{\Gamma(\alpha)} \frac{(g(y) - g(x))^{\alpha+1}}{(\alpha+1)}. \qquad (8.3.31)$$

We have proved that

$$\|f(x) - f(y) - (A_1(f))(x, y)(g(x) - g(y))\| \leq \qquad (8.3.32)$$

$$\frac{\lambda_2}{\Gamma(\alpha)} \frac{(g(y) - g(x))^{\alpha+1}}{(\alpha+1)}, \forall \, x, y \in [a, b] : y > x.$$

Conclusion 8.6 *Set* $\lambda := \max(\lambda_1, \lambda_2)$. *Then*

$$\|f(x) - f(y) - (A_1(f))(x, y)(g(x) - g(y))\| \leq$$

$$\frac{\lambda}{\Gamma(\alpha)} \frac{|g(x) - g(y)|^{\alpha+1}}{(\alpha+1)}, \forall \, x, y \in [a, b]. \qquad (8.3.33)$$

Notice that (8.3.33) is trivially true when $x = y$.
One may assume that

$$\frac{\lambda}{\Gamma(\alpha)} < 1. \qquad (8.3.34)$$

Now based on (8.3.20) and (8.3.33), we can apply our numerical methods presented in this chapter to solve $f(x) = 0$.

(II) In the next background again we use [5].
We need

Definition 8.7 ([5]) Let $\alpha > 0$, $\lceil \alpha \rceil = n$, $\lceil \cdot \rceil$ the ceiling of the number. Let $f \in C^n([a, b], X)$, where $[a, b] \subset \mathbb{R}$, and $(X, \|\cdot\|)$ is a Banach space. Let $g \in C^1([a, b])$, strictly increasing, such that $g^{-1} \in C^n([g(a), g(b)])$.

We define the right generalized g-fractional derivative X-valued of f of order α as follows:

$$\left(D_{b-;g}^\alpha f\right)(x) := \frac{(-1)^n}{\Gamma(n-\alpha)} \int_x^b (g(t) - g(x))^{n-\alpha-1} g'(t) \left(f \circ g^{-1}\right)^{(n)}(g(t)) \, dt, \qquad (8.3.35)$$

$\forall \, x \in [a, b]$. The last integral is of Bochner type.
If $\alpha \notin \mathbb{N}$, by [5], we have that $\left(D_{b-;g}^\alpha f\right) \in C([a, b], X)$.
We see that

$$J_{b-;g}^{n-\alpha} \left((-1)^n \left(f \circ g^{-1}\right)^{(n)} \circ g\right)(x) = \left(D_{b-;g}^\alpha f\right)(x), \, a \leq x \leq b. \qquad (8.3.36)$$

We set

$$D_{b-;g}^n f(x) := (-1)^n \left(\left(f \circ g^{-1}\right)^n \circ g\right)(x) \in C([a, b], X), n \in \mathbb{N}, \qquad (8.3.37)$$

$$D_{b-;g}^0 f(x) := f(x), \forall \, x \in [a, b].$$

When $g = id$, then

$$D_{b-;g}^\alpha f(x) = D_{b-;id}^\alpha f(x) = D_{b-}^\alpha f, \qquad (8.3.38)$$

the usual right X-valued Caputo fractional derivative, see [3].

We also need the Taylor's formula.

Theorem 8.8 ([5]) *Let* $\alpha > 0$, $n = \lceil \alpha \rceil$, *and* $f \in C^n([a, b], X)$, *where* $[a, b] \subset \mathbb{R}$ *and* $(X, \|\cdot\|)$ *is a Banach space. Let* $g \in C^1([a, b])$, *strictly increasing, such that* $g^{-1} \in C^n([g(a), g(b)])$, $a \leq x \leq b$. *Then*

$$f(x) = f(b) + \sum_{i=1}^{n-1} \frac{(g(x) - g(b))^i}{i!} \left(f \circ g^{-1} \right)^{(i)} (g(b)) +$$

$$\frac{1}{\Gamma(\alpha)} \int_x^b (g(t) - g(x))^{\alpha-1} g'(t) \left(D_{b-;g}^\alpha f \right)(t) \, dt =$$

$$f(b) + \sum_{i=1}^{n-1} \frac{(g(x) - g(b))^i}{i!} \left(f \circ g^{-1} \right)^{(i)} (g(b)) + \qquad (8.3.39)$$

$$\frac{1}{\Gamma(\alpha)} \int_{g(x)}^{g(b)} (z - g(x))^{\alpha-1} \left(\left(D_{b-;g}^\alpha f \right) \circ g^{-1} \right)(z) \, dz.$$

The remainder of (8.3.39) is a continuous function in $x \in [a, b]$.

Here we are going to operate more generally. We consider $f \in C^n([a, b], X)$. We define the following X-valued right g-fractional derivative of f of order α as follows:

$$\left(D_{y-;g}^\alpha f \right)(x) = \frac{(-1)^n}{\Gamma(n-\alpha)} \int_x^y (g(t) - g(x))^{n-\alpha-1} g'(t) \left(f \circ g^{-1} \right)^{(n)} (g(t)) \, dt,$$

$$(8.3.40)$$

$\forall x \in [a, y]$; where $y \in [a, b]$;

$$\left(D_{y-;g}^n f \right)(x) = (-1)^n \left(\left(f \circ g^{-1} \right)^{(n)} \circ g \right)(x), \forall x, y \in [a, b], \qquad (8.3.41)$$

$$\left(D_{y-;g}^0 f \right)(x) = f(x), \forall x \in [a, b]. \qquad (8.3.42)$$

For $\alpha > 0$, $\alpha \notin \mathbb{N}$, by convention we set that

$$\left(D_{y-;g}^\alpha f \right)(x) = 0, \text{ for } x > y, \forall x, y \in [a, b]. \qquad (8.3.43)$$

Similarly, we define

$$\left(D_{x-;g}^\alpha f \right)(y) = \frac{(-1)^n}{\Gamma(n-\alpha)} \int_y^x (g(t) - g(y))^{n-\alpha-1} g'(t) \left(f \circ g^{-1} \right)^{(n)} (g(t)) \, dt,$$

$$(8.3.44)$$

$\forall\, y \in [a, x]$, where $x \in [a, b]$;

$$\left(D_{x-;g}^{n} f\right)(y) = (-1)^{n} \left(\left(f \circ g^{-1}\right)^{(n)} \circ g\right)(y), \forall\, x, y \in [a, b], \qquad (8.3.45)$$

$$\left(D_{x-;g}^{0} f\right)(y) = f(y), \forall\, y \in [a, b]. \qquad (8.3.46)$$

For $\alpha > 0$, $\alpha \notin \mathbb{N}$, by convention we set that

$$\left(D_{x-;g}^{\alpha} f\right)(y) = 0, \text{ for } y > x, \ \forall\, x, y \in [a, b]. \qquad (8.3.47)$$

We get that

$$\left\|\left(D_{b-;g}^{\alpha} f\right)(x)\right\| \leq \frac{\left\|\left(f \circ g^{-1}\right)^{(n)} \circ g\right\|_{\infty,[a,b]}}{\Gamma(n - \alpha + 1)} (g(b) - g(x))^{n-\alpha} \leq \qquad (8.3.48)$$

$$\frac{\left\|\left(f \circ g^{-1}\right)^{(n)} \circ g\right\|_{\infty,[a,b]}}{\Gamma(n - \alpha + 1)} (g(b) - g(a))^{n-\alpha}, \ \forall\, x \in [a, b].$$

That is

$$\left(D_{b-;g}^{\alpha} f\right)(b) = 0, \qquad (8.3.49)$$

and

$$\left(D_{y-;g}^{\alpha} f\right)(y) = \left(D_{x-;g}^{\alpha} f\right)(x) = 0, \ \forall\, x, y \in [a, b]. \qquad (8.3.50)$$

Thus when $\alpha > 0$, $\alpha \notin \mathbb{N}$, both $D_{y-;g}^{\alpha} f$, $D_{x-;g}^{\alpha} f \in C([a, b], X)$, see [5].
Hence by Theorem 8.8 we obtain

$$f(x) - f(y) = \sum_{k=1}^{n-1} \frac{\left(f \circ g^{-1}\right)^{(k)}(g(y))}{k!} (g(x) - g(y))^{k} +$$

$$\frac{1}{\Gamma(\alpha)} \int_{x}^{y} (g(t) - g(x))^{\alpha-1} g'(t) \left(D_{y-;g}^{\alpha} f\right)(t)\, dt, \text{ all } a \leq x \leq y \leq b. \quad (8.3.51)$$

Also, we have

$$f(y) - f(x) = \sum_{k=1}^{n-1} \frac{\left(f \circ g^{-1}\right)^{(k)}(g(x))}{k!} (g(y) - g(x))^{k} +$$

$$\frac{1}{\Gamma(\alpha)} \int_{y}^{x} (g(t) - g(y))^{\alpha-1} g'(t) \left(D_{x-;g}^{\alpha} f\right)(t)\, dt, \text{ all } a \leq y \leq x \leq b. \quad (8.3.52)$$

We define also the following X-valued linear operator

$$(A_2(f))(x, y) :=$$

$$
\begin{cases}
\sum_{k=1}^{n-1} \frac{(f \circ g^{-1})^{(k)}(g(y))}{k!} (g(x) - g(y))^{k-1} - \left(D_{y-;g}^{\alpha} f\right)(x) \frac{(g(y)-g(x))^{\alpha-1}}{\Gamma(\alpha+1)}, & \text{for } x < y, \\
\sum_{k=1}^{n-1} \frac{(f \circ g^{-1})^{(k)}(g(x))}{k!} (g(y) - g(x))^{k-1} - \left(D_{x-;g}^{\alpha} f\right)(y) \frac{(g(x)-g(y))^{\alpha-1}}{\Gamma(\alpha+1)}, & \text{for } x > y, \\
f^{(n)}(x), & \text{when } x = y,
\end{cases}
$$

(8.3.53)

$\forall\, x, y \in [a, b];\ \alpha > 0, n = \lceil \alpha \rceil$.

We may assume that ([13], p. 3)

$$\|(A_2(f))(x, x) - (A_2(f))(y, y)\| = \left\| f^{(n)}(x) - f^{(n)}(y) \right\|$$ (8.3.54)

$$\leq \Phi^* |g(x) - g(y)|, \ \forall\, x, y \in [a, b];$$

where $\Phi^* > 0$.

We estimate and have

(i) case of $x < y$:

$$\| f(x) - f(y) - (A_2(f))(x, y)(g(x) - g(y))\| =$$

$$\left\| \frac{1}{\Gamma(\alpha)} \int_x^y (g(t) - g(x))^{\alpha-1} g'(t) \left(D_{y-;g}^{\alpha} f\right)(t)\, dt - \right.$$

$$\left. \left(D_{y-;g}^{\alpha} f\right)(x) \frac{(g(y) - g(x))^{\alpha}}{\Gamma(\alpha+1)} \right\|$$ (8.3.55)

(by [1] p. 426, Theorem 11.43)

$$= \frac{1}{\Gamma(\alpha)} \left\| \int_x^y (g(t) - g(x))^{\alpha-1} g'(t) \left(\left(D_{y-;g}^{\alpha} f\right)(t) - \left(D_{y-;g}^{\alpha} f\right)(x)\right) dt \right\|$$

(by [9])

$$\leq \frac{1}{\Gamma(\alpha)} \int_x^y (g(t) - g(x))^{\alpha-1} g'(t) \left\| \left(D_{y-;g}^{\alpha} f\right)(t) - \left(D_{y-;g}^{\alpha} f\right)(x) \right\| dt$$ (8.3.56)

(we assume that

$$\left\| \left(D_{y-;g}^{\alpha} f\right)(t) - \left(D_{y-;g}^{\alpha} f\right)(x) \right\| \leq \rho_1 |g(t) - g(x)|,$$ (8.3.57)

$\forall\, t, x, y \in [a, b] : y \geq t \geq x;\ \rho_1 > 0$)

$$\leq \frac{\rho_1}{\Gamma(\alpha)} \int_x^y (g(t) - g(x))^{\alpha-1} g'(t) (g(t) - g(x)) dt =$$

$$\frac{\rho_1}{\Gamma(\alpha)} \int_x^y (g(t) - g(x))^\alpha g'(t) dt = \frac{\rho_1}{\Gamma(\alpha)} \frac{(g(y) - g(x))^{\alpha+1}}{(\alpha+1)}. \tag{8.3.58}$$

We have proved that

$$\| f(x) - f(y) - (A_2(f))(x, y)(g(x) - g(y)) \| \leq$$

$$\frac{\rho_1}{\Gamma(\alpha)} \frac{(g(y) - g(x))^{\alpha+1}}{(\alpha+1)}, \tag{8.3.59}$$

$\forall\, x, y \in [a, b] : x < y$.

(ii) case of $x > y$:

$$\| f(x) - f(y) - (A_2(f))(x, y)(g(x) - g(y)) \| =$$

$$\| f(y) - f(x) - (A_2(f))(x, y)(g(y) - g(x)) \| = \tag{8.3.60}$$

$$\| f(y) - f(x) + (A_2(f))(x, y)(g(x) - g(y)) \| =$$

$$\left\| \frac{1}{\Gamma(\alpha)} \int_y^x (g(t) - g(y))^{\alpha-1} g'(t) \left(D_{x-;g}^\alpha f \right)(t)\, dt - \right.$$

$$\left. \left(D_{x-;g}^\alpha f \right)(y) \frac{(g(x) - g(y))^\alpha}{\Gamma(\alpha+1)} \right\| =$$

$$\frac{1}{\Gamma(\alpha)} \left\| \int_y^x (g(t) - g(y))^{\alpha-1} g'(t) \left(\left(D_{x-;g}^\alpha f \right)(t) - \left(D_{x-;g}^\alpha f \right)(y) \right) dt \right\| \leq$$

$$\tag{8.3.61}$$

$$\frac{1}{\Gamma(\alpha)} \int_y^x (g(t) - g(y))^{\alpha-1} g'(t) \left\| \left(D_{x-;g}^\alpha f \right)(t) - \left(D_{x-;g}^\alpha f \right)(y) \right\| dt \tag{8.3.62}$$

(we assume that

$$\left\| \left(D_{x-;g}^\alpha f \right)(t) - \left(D_{x-;g}^\alpha f \right)(y) \right\| \leq \rho_2 |g(t) - g(y)|, \tag{8.3.63}$$

$\forall\, t, y, x \in [a, b] : x \geq t \geq y; \rho_2 > 0$)

$$\leq \frac{\rho_2}{\Gamma(\alpha)} \int_y^x (g(t) - g(y))^{\alpha-1} g'(t) (g(t) - g(y)) dt =$$

$$\frac{\rho_2}{\Gamma(\alpha)} \int_y^x (g(t) - g(y))^\alpha g'(t) dt = \tag{8.3.64}$$

$$\frac{\rho_2}{\Gamma(\alpha)} \frac{(g(x) - g(y))^{\alpha+1}}{(\alpha+1)}. \tag{8.3.65}$$

We have proved that

$$\| f(x) - f(y) - (A_2(f))(x, y)(g(x) - g(y)) \| \le$$

$$\frac{\rho_2}{\Gamma(\alpha)} \frac{(g(x) - g(y))^{\alpha+1}}{(\alpha+1)}, \quad \forall\, x, y \in [a, b] : x > y. \tag{8.3.66}$$

Conclusion 8.9 *Set* $\rho := \max(\rho_1, \rho_2)$. *Then*

$$\| f(x) - f(y) - (A_2(f))(x, y)(g(x) - g(y)) \| \le$$

$$\frac{\rho}{\Gamma(\alpha)} \frac{|g(x) - g(y)|^{\alpha+1}}{(\alpha+1)}, \quad \forall\, x, y \in [a, b]. \tag{8.3.67}$$

Notice that (8.3.67) is trivially true when $x = y$.
One may assume that

$$\frac{\rho}{\Gamma(\alpha)} < 1. \tag{8.3.68}$$

Now based on (8.3.54) and (8.3.67), we can apply our numerical methods presented in this chapter to solve $f(x) = 0$.
In both fractional applications $\alpha + 1 \ge 2$, iff $\alpha \ge 1$.
Also some examples for g follow:

$$\begin{aligned}
g(x) &= e^x, x \in [a, b] \subset \mathbb{R}, \\
g(x) &= \sin x, \\
g(x) &= \tan x, \\
&\text{where } x \in \left[-\tfrac{\pi}{2} + \varepsilon, \tfrac{\pi}{2} - \varepsilon\right], \text{ where } \varepsilon > 0 \text{ small.}
\end{aligned} \tag{8.3.69}$$

References

1. C.D. Aliprantis, K.C. Border, *Infinite Dimensional Analysis* (Springer, New York, 2006)
2. S. Amat, S. Busquier, S. Plaza, Chaotic dynamics of a third-order Newton-type method. J. Math. Anal. Appl. **366**(1), 164–174 (2010)
3. G.A. Anastassiou, Strong right fractional calculus for Banach space valued functions. Rev. Proyecc. **36**(1), 149–186 (2017)
4. G.A. Anastassiou, A strong fractional calculus theory for Banach space valued functions. Nonlinear Funct. Anal. Appl. (Korea) (accepted for publication, 2017)
5. G.A. Anastassiou, *Principles of general fractional analysis for Banach space valued functions* (submitted for publication, 2017)
6. G.A. Anastassiou, I.K. Argyros, *Iterated convergence on Banach space valued functions of abstract g-fractional calculus* (submitted, 2017)

7. I.K. Argyros, A unifying local-semilocal convergence analysis and applications for two-point Newton-like methods in Banach space. J. Math. Anal. Appl. **298**, 374–397 (2004)
8. I.K. Argyros, A. Magréñan, *Iterative Methods and their Dynamics with Applications* (CRC Press, New York, 2017)
9. Bochner integral, *Encyclopedia of Mathematics*, http://www.encyclopediaofmath.org/index. php?title=Bochner_integral&oldid=38659
10. M. Edelstein, On fixed and periodic points under contractive mappings. J. London Math. Soc. **37**, 74–79 (1962)
11. J.A. Ezquerro, J.M. Gutierrez, M.A. Hernandez, N. Romero, M.J. Rubio, The Newton method: from Newton to Kantorovich (Spanish). Gac. R. Soc. Mat. Esp. **13**, 53–76 (2010)
12. L.V. Kantorovich, G.P. Akilov, *Functional Analysis in Normed Spaces* (Pergamon Press, New York, 1982)
13. G.E. Ladas, V. Lakshmikantham, *Differential Equations in Abstract Spaces* (Academic Press, New York, 1972)
14. A. Magréñan, A new tool to study real dynamics: the convergence plane. Appl. Math. Comput. **248**, 215–224 (2014)
15. J. Mikusinski, *The Bochner Integral.* (Academic Press, New York, 1978)
16. F.A. Potra, V. Pták, *Nondiscrete Induction and Iterative Processes* (Pitman Publication, London, 1984)
17. P.D. Proinov, New general convergence theory for iterative processes and its applications to Newton-Kantorovich type theorems. J. Complex. **26**, 3–42 (2010)
18. G.E. Shilov, *Elementary Functional Analysis* (Dover Publications Inc., New York, 1996)

Chapter 9
Generating Sequences for Solving in Abstract g-Fractional Calculus

The aim of this chapter is to utilize proper iterative methods for solving equations on Banach spaces. The differentiability of the operator involved is not assumed neither the convexity of its domain. Applications of the semi-local convergence are suggested including Banach space valued functions of fractional calculus, where all integrals are of Bochner-type. It follows [6].

9.1 Introduction

Sections 9.1–9.2 are prerequisites for Sect. 9.3.

Let B_1, B_2 stand for Banach spaces and let Ω stand for an open subset of B_1. Let also $U(z, \rho) := \{u \in B_1 : \|u - z\| < \rho\}$ and let $\overline{U}(z, \rho)$ stand for the closure of $U(z, \rho)$.

Many problems in Computational Sciences, Engineering, Mathematical Chemistry, Mathematical Physics, Mathematical Economics and other disciplines can written as

$$F(x) = 0 \tag{9.1.1}$$

using Mathematical Modeling [1–18], where $F : \Omega \to B_2$ is a continuous operator. The solution x^* of Eq. (9.1.1) is sought in closed form, but this is attainable only in special cases. That explains why most solution methods for such equations are usually iterative. There is a plethora of iterative methods for solving Eq. (9.1.1), more the [2, 7, 8, 10–14, 16, 17].

Newton's method [7, 8, 12, 16, 17]:

$$x_{n+1} = x_n - F'(x_n)^{-1} F(x_n). \tag{9.1.2}$$

Secant method:

$$x_{n+1} = x_n - [x_{n-1}, x_n; F]^{-1} F(x_n), \tag{9.1.3}$$

© Springer International Publishing AG 2018
G. A. Anastassiou and I. K. Argyros, *Functional Numerical Methods: Applications to Abstract Fractional Calculus*, Studies in Systems, Decision and Control 130, https://doi.org/10.1007/978-3-319-69526-6_9

where $[\cdot, \cdot; F]$ denotes a divided difference of order one on $\Omega \times \Omega$ [8, 16, 17].

Newton-like method:

$$x_{n+1} = x_n - E_n^{-1} F(x_n), \qquad (9.1.4)$$

where $E_n = E(F)(x_n)$ and $E : \Omega \to \mathcal{L}(B_1, B_2)$ the space of bounded linear operators from B_1 into B_2. Other methods can be found in [8, 12, 16, 17] and the references therein.

In the present study we consider the new method defined for each $n = 0, 1, 2, \ldots$ by

$$x_{n+1} = G(x_n)$$

$$G(x_{n+1}) = G(x_n) - A_n^{-1} F(x_n), \qquad (9.1.5)$$

where $x_0 \in \Omega$ is an initial point, $G : B_3 \to \Omega$ (B_3 a Banach space), $A_n = A(F)(x_{n+1}, x_n) = A(x_{n+1}, x_n)$ and $A : \Omega \times \Omega \to \mathcal{L}(B_1, B_2)$. Method (9.1.5) generates a sequence which we shall show converges to x^* under some Lipschitz-type conditions (to be precised in Sect. 9.2). Although method (9.1.5) (and Sect. 9.2) is of independent interest, it is nevertheless designed especially to be used in g-Abstract Fractional Calculus (to be precised in Sect. 9.3). As far as we know such iterative methods have not yet appeared in connection to solve equations in Abstract Fractional Calculus.

In this chapter we present the semi-local convergence of method (9.1.5) in Sect. 9.2. Some applications to Abstract g-Fractional Calculus are suggested in Sect. 9.3 on a certain Banach space valued functions, where all the integrals are of Bochner-type [9, 15].

9.2 Semi-local Convergence Analysis

We present the semi-local convergence analysis of method (9.1.5) using conditions (M):

(m_1) $F : \Omega \subset B_1 \to B_2$ is continuous, $G : B_3 \to \Omega$ is continuous and $A(x, y) \in \mathcal{L}(B_1, B_2)$ for each $(x, y) \in \Omega \times \Omega$.

(m_2) There exist $\beta > 0$ and $\Omega_0 \subset B_1$ such that $A(x, y)^{-1} \in \mathcal{L}(B_2, B_1)$ for each $(x, y) \in \Omega_0 \times \Omega_0$ and

$$\left\| A(x, y)^{-1} \right\| \leq \beta^{-1}.$$

Set $\Omega_1 = \Omega \cap \Omega_0$.

(m_3) There exists a continuous and nondecreasing function $\psi : [0, +\infty)^3 \to [0, +\infty)$ such that for each $x, y \in \Omega_1$

$$\| F(x) - F(y) - A(x, y)(G(x) - G(y)) \| \leq$$

$$\beta \psi (\|x - y\|, \|x - x_0\|, \|y - x_0\|) \|G(x) - G(y)\|.$$

(m_4) There exists a continuous and nondecreasing function $\psi_0 : [0, +\infty) \rightarrow [0, +\infty)$ such that for each $x \in \Omega_1$

$$\|G(x) - G(x_0)\| \leq \psi_0(\|x - x_0\|) \|x - x_0\|.$$

(m_5) For $x_0 \in \Omega_0$ and $x_1 = G(x_0) \in \Omega_0$ there exists $\eta \geq 0$ such that

$$\left\|A(x_1, x_0)^{-1} F(x_0)\right\| \leq \eta.$$

(m_6) There exists $s > 0$ such that

$$\psi(\eta, s, s) < 1,$$

$$\psi_0(s) < 1$$

and

$$\|G(x_0) - x_0\| \leq s \leq \frac{\eta}{1 - q_0},$$

where $q_0 = \psi(\eta, s, s)$.

(m_7) $\overline{U}(x_0, s) \subset \Omega$.

Next, we present the semi-local convergence analysis for method (9.1.5) using the conditions (M) and the preceding notation.

Theorem 9.1 *Assume that the conditions (M) hold. Then, sequence $\{x_n\}$ generated by method (9.1.5) starting at $x_0 \in \Omega$ is well defined in $U(x_0, s)$, remains in $U(x_0, s)$ for each $n = 0, 1, 2, \ldots$ and converges to a solution $x^* \in \overline{U}(x_0, s)$ of equation $F(x) = 0$. The limit point x^* is the unique solution of equation $F(x) = 0$ in $\overline{U}(x_0, s)$.*

Proof By the definition of s and (m_5), we have $x_1 \in U(x_0, s)$. The proof is based on mathematical induction on k. Suppose that $\|x_k - x_{k-1}\| \leq q_0^{k-1}\eta$ and $\|x_k - x_0\| \leq s$.
We get by (9.1.5), (m_2) − (m_5) in turn that

$$\|G(x_{k+1}) - G(x_k)\| = \left\|A_k^{-1} F(x_k)\right\| =$$

$$\left\|A_k^{-1}(F(x_k) - F(x_{k-1}) - A_{k-1}(G(x_k) - G(x_{k-1})))\right\|$$

$$\leq \left\|A_k^{-1}\right\| \|F(x_k) - F(x_{k-1}) - A_{k-1}(G(x_k) - G(x_{k-1}))\| \leq$$

$$\beta^{-1}\beta\psi(\|x_k - x_{k-1}\|, \|x_{k-1} - x_0\|, \|y_k - x_0\|) \|G(x_k) - G(x_{k-1})\| \leq$$

$$\psi(\eta, s, s) \|G(x_k) - G(x_{k-1})\| = q_0 \|G(x_k) - G(x_{k-1})\| \leq q_0^k \|x_1 - x_0\| \leq q_0^k\eta$$

$$(9.2.1)$$

and by (m_6)

$$\|x_{k+1} - x_0\| = \|G(x_k) - x_0\| \leq \|G(x_k) - G(x_0)\| + \|G(x_0) - x_0\|$$

$$\leq \psi_0(\|x_k - x_0\|)\|x_k - x_0\| + \|G(x_0) - x_0\|$$

$$\leq \psi_0(s)s + \|G(x_0) - x_0\| \leq s.$$

The induction is completed. Moreover, we have by (9.2.1) that for $m = 0, 1, 2, \ldots$

$$\|x_{k+m} - x_k\| \leq \frac{1 - q_0^m}{1 - q_0} q_0^k \eta.$$

It follows from the preceding inequation that sequence $\{G(x_k)\}$ is complete in a Banach space B_1 and as such it converges to some $x^* \in \overline{U}(x_0, s)$ (since $\overline{U}(x_0, s)$ is a closed ball). By letting $k \to +\infty$ in (9.2.1) we get $F(x^*) = 0$. We also get by (9.1.5) that $G(x^*) = x^*$. To show the uniqueness part, let $x^{**} \in U(x_0, s)$ be a solution of equation $F(x) = 0$ and $G(x^{**}) = x^{**}$. By using (9.1.5), we obtain in turn that

$$\left\|x^{**} - G(x_{k+1})\right\| = \left\|x^{**} - G(x_k) + A_k^{-1} F(x_k) - A_k^{-1} F(x^{**})\right\| \leq$$

$$\left\|A_k^{-1}\right\| \left\|F(x^{**}) - F(x_k) - A_k(G(x^{**}) - G(x_k))\right\| \leq$$

$$\beta^{-1} \beta \psi_0 \left(\left\|x^{**} - x_k\right\|, \|x_{k+1} - x_0\|, \|x_k - x_0\|\right) \left\|G(x^{**}) - G(x_k)\right\| \leq$$

$$q_0 \left\|G(x^{**}) - G(x_k)\right\| \leq q_0^{k+1} \left\|x^{**} - x_0\right\|,$$

so $\lim_{k \to +\infty} x_k = x^{**}$. We have shown that $\lim_{k \to +\infty} x_k = x^*$, so $x^* = x^{**}$. ∎

Remark 9.2 (1) Condition (m_2) can become part of condition (m_3) by considering
$(m_3)'$ There exists a continuous and nondecreasing function $\varphi : [0, +\infty)^3 \to [0, +\infty)$ such that for each $x, y \in \Omega_1$

$$\left\|A(x, y)^{-1} [F(x) - F(y) - A(x, y)(G(x) - G(y))]\right\| \leq$$

$$\varphi(\|x - y\|, \|x - x_0\|, \|y - x_0\|) \|G(x) - G(y)\|.$$

Notice that

$$\varphi(u_1, u_2, u_3) \leq \psi(u_1, u_2, u_3)$$

for each $u_1 \geq 0$, $u_2 \geq 0$ and $u_3 \geq 0$. Similarly, a function φ_1 can replace ψ_1 for the uniqueness of the solution part. These replacements are of Mysovskii-type

[7, 12, 16] and influence the weaking of the convergence criterion in (m_6), error bounds and the precision of s.

(2) Suppose that there exist $\beta > 0$, $\beta_1 > 0$ and $L \in \mathcal{L}(B_1, B_2)$ with $L^{-1} \in \mathcal{L}(B_2, B_1)$ such that

$$\left\| L^{-1} \right\| \leq \beta^{-1}$$

$$\left\| A(x, y) - L \right\| \leq \beta_1$$

and

$$\beta_2 := \beta^{-1}\beta_1 < 1.$$

Then, it follows from the Banach lemma on invertible operators [12], and

$$\left\| L^{-1} \right\| \left\| A(x, y) - L \right\| \leq \beta^{-1}\beta_1 = \beta_2 < 1$$

that $A(x, y)^{-1} \in \mathcal{L}(B_2, B_1)$. Let $\beta = \frac{\beta^{-1}}{1-\beta_2}$. Then, under these replacements, condition (m_2) is implied, therefore it can be dropped from the conditions (M).

Remark 9.3 Sect. 9.2 has an interest independent of Sect. 9.3. It is worth noticing that the results especially of Theorem 9.1 can apply in Abstract g-Fractional Calculus as illustrated in Sect. 9.3. By specializing function ψ, we can apply the results of say Theorem 9.1 in the examples suggested in Sect. 9.3. In particular for (9.3.21), we choose for $u_1 \geq 0$, $u_2 \geq 0$, $u_3 \geq 0$

$$\psi(u_1, u_2, u_3) = \frac{\lambda \mu_1^\nu}{\beta \Gamma(\nu)(\nu+1)},$$

if $|g(x) - g(y)| \leq \mu_1$ for each $x, y \in [a, b]$;

$$\psi(u_1, u_2, u_3) = \frac{\lambda \mu_2^\nu}{\beta \Gamma(\nu)(\nu+1)},$$

if $|g(x) - g(y)| \leq \xi_2 \|x - y\|$ for each $x, y \in [a, b]$ and $\mu_2 = \xi_2 |b - a|$;

$$\psi(u_1, u_2, u_3) = \frac{\lambda \mu_3^\nu}{\beta \Gamma(\nu)(\nu+1)},$$

if $|g(x)| \leq \xi_3$ for each $x, y \in [a, b]$ and $\mu_3 = 2\xi_3$, where λ, ν and F are defined in Sect. 9.3. Other choices of function ψ are also possible.

Notice that with these choices of function ψ and $f = F$ and $g = G$, crucial condition (m_3) is satisfied, which justifies our definition of method (9.1.5). We can provide similar choices for the other examples of Sect. 9.3.

9.3   Applications to X-valued g-Fractional Calculus of Canavati Type

Here we deal with Banach space $(X, \|\cdot\|)$ valued functions f of real domain $[a, b]$. All integrals here are of Bochner-type, see [15]. The derivatives of f are defined similarly to numerical ones, see [18], pp. 83–86 and p. 93.

Here both needed backgrounds come from [5].

Let $\nu > 1$, $\nu \notin \mathbb{N}$, with integral part $[\nu] = n \in \mathbb{N}$. Let $g : [a, b] \to \mathbb{R}$ be a strictly increasing function, such that $g \in C^1([a, b])$, $g^{-1} \in C^n([g(a), g(b)])$, and let $f \in C^n([a, b], X)$. It clear then we obtain that $(f \circ g^{-1}) \in C^n([g(a), g(b)], X)$. Let $\alpha := \nu - [\nu] = \nu - n$ $(0 < \alpha < 1)$.

(I) See [5]. Let $h \in C([g(a), g(b)], X)$, we define the X-valued left Riemann-Liouville fractional integral as

$$\left(J_\nu^{z_0} h\right)(z) := \frac{1}{\Gamma(\nu)} \int_{z_0}^z (z - t)^{\nu-1} h(t)\, dt, \tag{9.3.1}$$

for $g(a) \leq z_0 \leq z \leq g(b)$, where Γ is the gamma function.

We define the subspace $C_{g(x)}^\nu([g(a), g(b)], X)$ of $C^n([g(a), g(b)], X)$, where $x \in [a, b]$:

$$C_{g(x)}^\nu([g(a), g(b)], X) :=$$

$$\left\{ h \in C^n([g(a), g(b)], X) : J_{1-\alpha}^{g(x)} h^{(n)} \in C^1([g(x), g(b)], X) \right\}. \tag{9.3.2}$$

So let $h \in C_{g(x)}^\nu([g(a), g(b)], X)$; we define the X-valued left g-generalized fractional derivative of h of order ν, of Canavati type, over $[g(x), g(b)]$ as

$$D_{g(x)}^\nu h := \left(J_{1-\alpha}^{g(x)} h^{(n)} \right)'. \tag{9.3.3}$$

Clearly, for $h \in C_{g(x)}^\nu([g(a), g(b)], X)$, there exists

$$\left(D_{g(x)}^\nu h\right)(z) = \frac{1}{\Gamma(1-\alpha)} \frac{d}{dz} \int_{g(x)}^z (z - t)^{-\alpha} h^{(n)}(t)\, dt, \tag{9.3.4}$$

for all $g(x) \leq z \leq g(b)$.

In particular, when $f \circ g^{-1} \in C_{g(x)}^\nu([g(a), g(b)], X)$ we have that

$$\left(D_{g(x)}^\nu \left(f \circ g^{-1}\right)\right)(z) = \frac{1}{\Gamma(1-\alpha)} \frac{d}{dz} \int_{g(x)}^z (z - t)^{-\alpha} \left(f \circ g^{-1}\right)^{(n)}(t)\, dt, \tag{9.3.5}$$

for all $z : g(x) \leq z \leq g(b)$.

We have that $D_{g(x)}^n \left(f \circ g^{-1}\right) = \left(f \circ g^{-1}\right)^{(n)}$ and $D_{g(x)}^0 \left(f \circ g^{-1}\right) = f \circ g^{-1}$.

From [5] we have for $\left(f \circ g^{-1}\right) \in C^{\nu}_{g(x)}\left([g(a), g(b)], X\right)$, where $x \in [a, b]$, (X-valued left fractional Taylor's formula) that

$$f(y) - f(x) = \sum_{k=1}^{n-1} \frac{\left(f \circ g^{-1}\right)^{(k)}(g(x))}{k!} (g(y) - g(x))^k + \qquad (9.3.6)$$

$$\frac{1}{\Gamma(\nu)} \int_{g(x)}^{g(y)} (g(y) - t)^{\nu-1} \left(D^{\nu}_{g(x)} \left(f \circ g^{-1}\right)\right)(t)\, dt, \quad \text{for all } y \in [a, b] : y \geq x.$$

Alternatively, for $\left(f \circ g^{-1}\right) \in C^{\nu}_{g(y)}\left([g(a), g(b)], X\right)$, where $y \in [a, b]$, we can write (again X-valued left fractional Taylor's formula) that:

$$f(x) - f(y) = \sum_{k=1}^{n-1} \frac{\left(f \circ g^{-1}\right)^{(k)}(g(y))}{k!} (g(x) - g(y))^k + \qquad (9.3.7)$$

$$\frac{1}{\Gamma(\nu)} \int_{g(y)}^{g(x)} (g(x) - t)^{\nu-1} \left(D^{\nu}_{g(y)} \left(f \circ g^{-1}\right)\right)(t)\, dt, \quad \text{for all } x \in [a, b] : x \geq y.$$

Here we consider $f \in C^n([a, b], X)$, such that $\left(f \circ g^{-1}\right) \in C^{\nu}_{g(x)}([g(a), g(b)], X)$, for every $x \in [a, b]$; which is the same as $\left(f \circ g^{-1}\right) \in C^{\nu}_{g(y)}([g(a), g(b)], X)$, for every $y \in [a, b]$ (i.e. exchange roles of x and y); we write that as $\left(f \circ g^{-1}\right) \in C^{\nu}_{g+}([g(a), g(b)], X)$.

We have that

$$\left(D^{\nu}_{g(y)} \left(f \circ g^{-1}\right)\right)(z) = \frac{1}{\Gamma(1-\alpha)} \frac{d}{dz} \int_{g(y)}^{z} (z - t)^{-\alpha} \left(f \circ g^{-1}\right)^{(n)}(t)\, dt, \quad (9.3.8)$$

for all $z : g(y) \leq z \leq g(b)$.

So here we work with $f \in C^n([a, b], X)$, such that $\left(f \circ g^{-1}\right) \in C^{\nu}_{g+}([g(a), g(b)], X)$.

We define the X-valued left linear fractional operator

$$(A_1(f))(x, y) := \begin{cases} \sum_{k=1}^{n-1} \frac{(f \circ g^{-1})^{(k)}(g(x))}{k!} (g(y) - g(x))^{k-1} + \\ \left(D^{\nu}_{g(x)} \left(f \circ g^{-1}\right)\right)(g(y)) \frac{(g(y) - g(x))^{\nu-1}}{\Gamma(\nu+1)}, \quad y > x, \\[2ex] \sum_{k=1}^{n-1} \frac{(f \circ g^{-1})^{(k)}(g(y))}{k!} (g(x) - g(y))^{k-1} + \\ \left(D^{\nu}_{g(y)} \left(f \circ g^{-1}\right)\right)(g(x)) \frac{(g(x) - g(y))^{\nu-1}}{\Gamma(\nu+1)}, \quad x > y, \\[2ex] f^{(n)}(x), \quad x = y. \end{cases} \qquad (9.3.9)$$

We may assume that (see [13], p. 3)

$$\|(A_1(f))(x,x) - (A_1(f))(y,y)\| = \|f^{(n)}(x) - f^{(n)}(y)\| =$$

$$\left\|\left(f^{(n)} \circ g^{-1}\right)(g(x)) - \left(f^{(n)} \circ g^{-1}\right)(g(y))\right\| \leq \Phi|g(x) - g(y)|, \qquad (9.3.10)$$

where $\Phi > 0$; for any $x, y \in [a, b]$.

We make the following estimations:

(i) case of $y > x$: We have that

$$\|f(y) - f(x) - (A_1(f))(x,y)(g(y) - g(x))\| =$$

$$\left\| \frac{1}{\Gamma(\nu)} \int_{g(x)}^{g(y)} (g(y) - t)^{\nu-1} \left(D_{g(x)}^{\nu} \left(f \circ g^{-1}\right)\right)(t) \, dt - \right.$$

$$\left. \left(D_{g(x)}^{\nu}\left(f \circ g^{-1}\right)\right)(g(y)) \frac{(g(y) - g(x))^{\nu}}{\Gamma(\nu+1)} \right\|$$

(by [1], p. 426, Theorem 11.43)

$$= \frac{1}{\Gamma(\nu)} \left\| \int_{g(x)}^{g(y)} (g(y) - t)^{\nu-1} \left(\left(D_{g(x)}^{\nu}\left(f \circ g^{-1}\right)\right)(t) - \left(D_{g(x)}^{\nu}\left(f \circ g^{-1}\right)\right)(g(y)) \right) dt \right\|$$
$$(9.3.11)$$

(by [9])

$$\leq \frac{1}{\Gamma(\nu)} \int_{g(x)}^{g(y)} (g(y) - t)^{\nu-1} \left\| \left(D_{g(x)}^{\nu}\left(f \circ g^{-1}\right)\right)(t) - \left(D_{g(x)}^{\nu}\left(f \circ g^{-1}\right)\right)(g(y)) \right\| dt$$

(we assume here that

$$\left\|\left(D_{g(x)}^{\nu}\left(f \circ g^{-1}\right)\right)(t) - \left(D_{g(x)}^{\nu}\left(f \circ g^{-1}\right)\right)(g(y))\right\| \leq \lambda_1 |t - g(y)|, \quad (9.3.12)$$

for every $t, g(y), g(x) \in [g(a), g(b)]$ such that $g(y) \geq t \geq g(x)$; $\lambda_1 > 0$)

$$\leq \frac{\lambda_1}{\Gamma(\nu)} \int_{g(x)}^{g(y)} (g(y) - t)^{\nu-1} (g(y) - t) \, dt = \qquad (9.3.13)$$

$$\frac{\lambda_1}{\Gamma(\nu)} \int_{g(x)}^{g(y)} (g(y) - t)^{\nu} \, dt = \frac{\lambda_1}{\Gamma(\nu)} \frac{(g(y) - g(x))^{\nu+1}}{(\nu+1)}. \qquad (9.3.14)$$

We have proved that

$$\|f(y) - f(x) - (A_1(f))(x,y)(g(y) - g(x))\| \leq \frac{\lambda_1}{\Gamma(\nu)} \frac{(g(y) - g(x))^{\nu+1}}{(\nu+1)},$$
$$(9.3.15)$$

for all $x, y \in [a, b] : y > x$.

(ii) Case of $x > y$: We observe that

$$\| f(y) - f(x) - (A_1(f))(x, y)(g(y) - g(x)) \| =$$

$$\| f(x) - f(y) - (A_1(f))(x, y)(g(x) - g(y)) \| =$$

$$\left\| \frac{1}{\Gamma(\nu)} \int_{g(y)}^{g(x)} (g(x) - t)^{\nu-1} \left(D_{g(y)}^{\nu}(f \circ g^{-1}) \right)(t)\, dt - \right.$$

$$\left. \left(D_{g(y)}^{\nu}(f \circ g^{-1}) \right)(g(x)) \frac{(g(x) - g(y))^{\nu}}{\Gamma(\nu+1)} \right\| = \qquad (9.3.16)$$

$$\frac{1}{\Gamma(\nu)} \left\| \int_{g(y)}^{g(x)} (g(x) - t)^{\nu-1} \left(\left(D_{g(y)}^{\nu}(f \circ g^{-1}) \right)(t) - \left(D_{g(y)}^{\nu}(f \circ g^{-1}) \right)(g(x)) \right) dt \right\|$$

$$\leq \frac{1}{\Gamma(\nu)} \int_{g(y)}^{g(x)} (g(x) - t)^{\nu-1} \left\| \left(D_{g(y)}^{\nu}(f \circ g^{-1}) \right)(t) - \left(D_{g(y)}^{\nu}(f \circ g^{-1}) \right)(g(x)) \right\| dt$$

$$\qquad (9.3.17)$$

(we assume that

$$\left\| \left(D_{g(y)}^{\nu}(f \circ g^{-1}) \right)(t) - \left(D_{g(y)}^{\nu}(f \circ g^{-1}) \right)(g(x)) \right\| \leq \lambda_2 \, |t - g(x)| , \quad (9.3.18)$$

for all $t, g(x), g(y) \in [g(a), g(b)]$ such that $g(x) \geq t \geq g(y)$; $\lambda_2 > 0$)

$$\leq \frac{\lambda_2}{\Gamma(\nu)} \int_{g(y)}^{g(x)} (g(x) - t)^{\nu-1} (g(x) - t)\, dt = \qquad (9.3.19)$$

$$\frac{\lambda_2}{\Gamma(\nu)} \int_{g(y)}^{g(x)} (g(x) - t)^{\nu}\, dt = \frac{\lambda_2}{\Gamma(\nu)} \frac{(g(x) - g(y))^{\nu+1}}{(\nu+1)} .$$

We have proved that

$$\| f(y) - f(x) - (A_1(f))(x, y)(g(y) - g(x)) \| \leq \frac{\lambda_2}{\Gamma(\nu)} \frac{(g(x) - g(y))^{\nu+1}}{(\nu+1)} ,$$

$$\qquad (9.3.20)$$

for any $x, y \in [a, b] : x > y$.

Conclusion 9.4 *Set $\lambda := \max(\lambda_1, \lambda_2)$. Then*

$$\| f(y) - f(x) - (A_1(f))(x, y)(g(y) - g(x)) \| \leq \frac{\lambda}{\Gamma(\nu)} \frac{|g(y) - g(x)|^{\nu+1}}{(\nu+1)} ,$$

$$\qquad (9.3.21)$$

$\forall\, x, y \in [a, b]$ *(the case of $x = y$ is trivially true).*

We may choose that $\frac{\lambda}{\Gamma(\nu)} < 1$.

Also we notice here that $\nu + 1 > 2$.

(II) See [5] again. Let $h \in C([g(a), g(b)], X)$, we define the X-valued right Riemann-Liouville fractional integral as

$$\left(J_{z_0-}^\nu h\right)(z) := \frac{1}{\Gamma(\nu)} \int_z^{z_0} (t-z)^{\nu-1} h(t) \, dt, \tag{9.3.22}$$

for $g(a) \le z \le z_0 \le g(b)$.

We define the subspace $C_{g(x)-}^\nu([g(a), g(b)], X)$ of $C^n([g(a), g(b)], X)$, where $x \in [a, b]$:

$$C_{g(x)-}^\nu([g(a), g(b)], X) :=$$

$$\left\{ h \in C^n([g(a), g(b)], X) : J_{g(x)-}^{1-\alpha} h^{(n)} \in C^1([g(a), g(x)], X) \right\}. \tag{9.3.23}$$

So let $h \in C_{g(x)-}^\nu([g(a), g(b)], X)$; we define the X-valued right g-generalized fractional derivative of h of order ν, of Canavati type, over $[g(a), g(x)]$ as

$$D_{g(x)-}^\nu h := (-1)^{n-1} \left(J_{g(x)-}^{1-\alpha} h^{(n)}\right)'. \tag{9.3.24}$$

Clearly, for $h \in C_{g(x)-}^\nu([g(a), g(b)], X)$, there exists

$$\left(D_{g(x)-}^\nu h\right)(z) = \frac{(-1)^{n-1}}{\Gamma(1-\alpha)} \frac{d}{dz} \int_z^{g(x)} (t-z)^{-\alpha} h^{(n)}(t) \, dt, \tag{9.3.25}$$

for all $g(a) \le z \le g(x) \le g(b)$.

In particular, when $f \circ g^{-1} \in C_{g(x)-}^\nu([g(a), g(b)], X)$ we have that

$$\left(D_{g(x)-}^\nu \left(f \circ g^{-1}\right)\right)(z) = \frac{(-1)^{n-1}}{\Gamma(1-\alpha)} \frac{d}{dz} \int_z^{g(x)} (t-z)^{-\alpha} \left(f \circ g^{-1}\right)^{(n)}(t) \, dt, \tag{9.3.26}$$

for all $g(a) \le z \le g(x) \le g(b)$.

We get that

$$\left(D_{g(x)-}^n \left(f \circ g^{-1}\right)\right)(z) = (-1)^n \left(f \circ g^{-1}\right)^{(n)}(z), \tag{9.3.27}$$

and

$$\left(D_{g(x)-}^0 \left(f \circ g^{-1}\right)\right)(z) = \left(f \circ g^{-1}\right)(z), \tag{9.3.28}$$

for all $z \in [g(a), g(x)]$, see [5].

From [5] we have, for $\left(f \circ g^{-1}\right) \in C_{g(x)-}^\nu([g(a), g(b)], X)$, where $x \in [a, b]$, $\nu \ge 1$ (X-valued right fractional Taylor's formula) that:

$$f(y) - f(x) = \sum_{k=1}^{n-1} \frac{\left(f \circ g^{-1}\right)^{(k)}(g(x))}{k!} (g(y) - g(x))^k +$$

$$\frac{1}{\Gamma(\nu)} \int_{g(y)}^{g(x)} (t - g(y))^{\nu-1} \left(D_{g(x)-}^{\nu} \left(f \circ g^{-1}\right)\right)(t)\, dt, \text{ all } a \le y \le x. \quad (9.3.29)$$

Alternatively, for $\left(f \circ g^{-1}\right) \in C_{g(y)-}^{\nu}\left([g(a), g(b)], X\right)$, where $y \in [a, b], \nu \ge 1$ (again X-valued right fractional Taylor's formula) that:

$$f(x) - f(y) = \sum_{k=1}^{n-1} \frac{\left(f \circ g^{-1}\right)^{(k)}(g(y))}{k!} (g(x) - g(y))^k +$$

$$\frac{1}{\Gamma(\nu)} \int_{g(x)}^{g(y)} (t - g(x))^{\nu-1} \left(D_{g(y)-}^{\nu} \left(f \circ g^{-1}\right)\right)(t)\, dt, \text{ all } a \le x \le y. \quad (9.3.30)$$

Here we consider $f \in C^n([a, b], X)$, such that $\left(f \circ g^{-1}\right) \in C_{g(x)-}^{\nu}([g(a), g(b)], X)$, for every $x \in [a, b]$; which is the same as $\left(f \circ g^{-1}\right) \in C_{g(y)-}^{\nu}([g(a), g(b)], X)$, for every $y \in [a, b]$; (i.e. exchange roles of x and y) we write that as $\left(f \circ g^{-1}\right) \in C_{g-}^{\nu}([g(a), g(b)], X)$.

We have that

$$\left(D_{g(y)-}^{\nu} \left(f \circ g^{-1}\right)\right)(z) = \frac{(-1)^{n-1}}{\Gamma(1-\alpha)} \frac{d}{dz} \int_{z}^{g(y)} (t - z)^{-\alpha} \left(f \circ g^{-1}\right)^{(n)}(t)\, dt,$$

$$(9.3.31)$$

for all $g(a) \le z \le g(y) \le g(b)$.

So here we work with $f \in C^n([a, b], X)$, such that $\left(f \circ g^{-1}\right) \in C_{g-}^{\nu}([g(a), g(b)], X)$.

We define the X-valued right linear fractional operator

$$(A_2(f))(x, y) := \begin{cases} \sum_{k=1}^{n-1} \frac{(f \circ g^{-1})^{(k)}(g(x))}{k!} (g(y) - g(x))^{k-1} - \\ \left(D_{g(x)-}^{\nu}\left(f \circ g^{-1}\right)\right)(g(y)) \frac{(g(x)-g(y))^{\nu-1}}{\Gamma(\nu+1)}, & x > y, \\[4mm] \sum_{k=1}^{n-1} \frac{(f \circ g^{-1})^{(k)}(g(y))}{k!} (g(x) - g(y))^{k-1} - \\ \left(D_{g(y)-}^{\nu}\left(f \circ g^{-1}\right)\right)(g(x)) \frac{(g(y)-g(x))^{\nu-1}}{\Gamma(\nu+1)}, & y > x, \\[4mm] f^{(n)}(x), & x = y. \end{cases} \quad (9.3.32)$$

We may assume that ([13], p. 3)

$$\|(A_2(f))(x, x) - (A_2(f))(y, y)\| = \left\|f^{(n)}(x) - f^{(n)}(y)\right\| \le \Phi^* |g(x) - g(y)|,$$

$$(9.3.33)$$

where $\Phi^* > 0$; for any $x, y \in [a, b]$.

We make the following estimations:

(i) case of $x > y$: We have that

$$\| f(x) - f(y) - (A_2(f))(x, y)(g(x) - g(y)) \| =$$

$$\| f(y) - f(x) - (A_2(f))(x, y)(g(y) - g(x)) \| = \qquad (9.3.34)$$

$$\| f(y) - f(x) + (A_2(f))(x, y)(g(x) - g(y)) \| =$$

$$\left\| \frac{1}{\Gamma(\nu)} \int_{g(y)}^{g(x)} (t - g(y))^{\nu-1} \left(D_{g(x)-}^{\nu} \left(f \circ g^{-1} \right) \right)(t) \, dt - \right.$$

$$\left. \left(D_{g(x)-}^{\nu} \left(f \circ g^{-1} \right) \right)(g(y)) \frac{(g(x) - g(y))^{\nu}}{\Gamma(\nu + 1)} \right\| \qquad (9.3.35)$$

(by [1], p. 426, Theorem 11.43)

$$= \frac{1}{\Gamma(\nu)} \left\| \int_{g(y)}^{g(x)} (t - g(y))^{\nu-1} \left(\left(D_{g(x)-}^{\nu} \left(f \circ g^{-1} \right) \right)(t) - \left(D_{g(x)-}^{\nu} \left(f \circ g^{-1} \right) \right) \right) \right.$$

$$\left. (g(y)) \, dt \right\|$$

(by [9])

$$\leq \frac{1}{\Gamma(\nu)} \int_{g(y)}^{g(x)} (t - g(y))^{\nu-1} \left\| \left(D_{g(x)-}^{\nu} \left(f \circ g^{-1} \right) \right)(t) - \left(D_{g(x)-}^{\nu} \left(f \circ g^{-1} \right) \right)(g(y)) \right\| dt$$

$$\qquad (9.3.36)$$

(we assume here that

$$\left\| \left(D_{g(x)-}^{\nu} \left(f \circ g^{-1} \right) \right)(t) - \left(D_{g(x)-}^{\nu} \left(f \circ g^{-1} \right) \right)(g(y)) \right\| \leq \rho_1 |t - g(y)|, \quad (9.3.37)$$

for every $t, g(y), g(x) \in [g(a), g(b)]$ such that $g(x) \geq t \geq g(y)$; $\rho_1 > 0$)

$$\leq \frac{\rho_1}{\Gamma(\nu)} \int_{g(y)}^{g(x)} (t - g(y))^{\nu-1} (t - g(y)) \, dt =$$

$$\frac{\rho_1}{\Gamma(\nu)} \int_{g(y)}^{g(x)} (t - g(y))^{\nu} \, dt = \frac{\rho_1}{\Gamma(\nu)} \frac{(g(x) - g(y))^{\nu+1}}{(\nu + 1)}. \qquad (9.3.38)$$

We have proved that

$$\|f(x) - f(y) - (A_2(f))(x,y)(g(x) - g(y))\| \leq \frac{\rho_1}{\Gamma(\nu)} \frac{(g(x) - g(y))^{\nu+1}}{(\nu+1)},$$

(9.3.39)

$\forall\, x, y \in [a,b] : x > y$.

(ii) Case of $x < y$: We have that

$$\|f(x) - f(y) - (A_2(f))(x,y)(g(x) - g(y))\| =$$

$$\|f(x) - f(y) + (A_2(f))(x,y)(g(y) - g(x))\| = \qquad (9.3.40)$$

$$\left\| \frac{1}{\Gamma(\nu)} \int_{g(x)}^{g(y)} (t - g(x))^{\nu-1} \left(D_{g(y)-}^{\nu}(f \circ g^{-1})\right)(t)\, dt - \right.$$

$$\left. \left(D_{g(y)-}^{\nu}(f \circ g^{-1})\right)(g(x)) \frac{(g(y) - g(x))^{\nu}}{\Gamma(\nu+1)} \right\| =$$

$$\frac{1}{\Gamma(\nu)} \left\| \int_{g(x)}^{g(y)} (t - g(x))^{\nu-1} \left(\left(D_{g(y)-}^{\nu}(f \circ g^{-1})\right)(t) - \left(D_{g(y)-}^{\nu}(f \circ g^{-1})\right)(g(x)) \right) dt \right\|$$

$$\leq \frac{1}{\Gamma(\nu)} \int_{g(x)}^{g(y)} (t - g(x))^{\nu-1} \left\| \left(D_{g(y)-}^{\nu}(f \circ g^{-1})\right)(t) - \left(D_{g(y)-}^{\nu}(f \circ g^{-1})\right)(g(x)) \right\| dt$$

(9.3.41)

(we assume that

$$\left\| \left(D_{g(y)-}^{\nu}(f \circ g^{-1})\right)(t) - \left(D_{g(y)-}^{\nu}(f \circ g^{-1})\right)(g(x)) \right\| \leq \rho_2 |t - g(x)|, \quad (9.3.42)$$

for any $t, g(x), g(y) \in [g(a), g(b)] : g(y) \geq t \geq g(x)$; $\rho_2 > 0$)

$$\leq \frac{\rho_2}{\Gamma(\nu)} \int_{g(x)}^{g(y)} (t - g(x))^{\nu-1} (t - g(x))\, dt =$$

$$\frac{\rho_2}{\Gamma(\nu)} \int_{g(x)}^{g(y)} (t - g(x))^{\nu}\, dt = \qquad (9.3.43)$$

$$\frac{\rho_2}{\Gamma(\nu)} \frac{(g(y) - g(x))^{\nu+1}}{(\nu+1)}. \qquad (9.3.44)$$

We have proved that

$$\|f(x) - f(y) - (A_2(f))(x,y)(g(x) - g(y))\| \leq \frac{\rho_2}{\Gamma(\nu)} \frac{(g(y) - g(x))^{\nu+1}}{(\nu+1)},$$

(9.3.45)

$\forall\, x, y \in [a,b] : x < y$.

Conclusion 9.5 *Set $\rho := \max(\rho_1, \rho_2)$. Then*

$$\|f(x) - f(y) - (A_2(f))(x,y)(g(x) - g(y))\| \le \frac{\rho}{\Gamma(\nu)} \frac{|g(x) - g(y)|^{\nu+1}}{(\nu+1)},$$
(9.3.46)

$\forall\, x, y \in [a, b]$ *((9.3.46) is trivially true when $x = y$).*

One may choose $\frac{\rho}{\Gamma(\nu)} < 1$.
Here again $\nu + 1 > 2$.

Conclusion 9.6 *Based on (9.3.10) and (9.3.21) of (I), and based on (9.3.33) and (9.3.46) of (II), using our numerical results presented earlier, we can solve numerically $f(x) = 0$.*

Some examples for g follow:

$$g(x) = e^x, \, x \in [a, b] \subset \mathbb{R},$$
$$g(x) = \sin x,$$
$$g(x) = \tan x,$$
$$\text{where } x \in \left[-\frac{\pi}{2} + \varepsilon, \frac{\pi}{2} - \varepsilon\right], \text{ with } \varepsilon > 0 \text{ small.}$$

References

1. C.D. Aliprantis, K.C. Border, *Infinite Dimensional Analysis* (Springer, New York, 2006)
2. S. Amat, S. Busquier, S. Plaza, Chaotic dynamics of a third-order Newton-type method. J. Math. Anal. Appl. **366**(1), 164–174 (2010)
3. G.A. Anastassiou, Strong right fractional calculus for banach space valued functions. Rev Proyecc. **36**(1), 149–186 (2017)
4. G.A. Anastassiou, *A strong Fractional Calculus Theory for Banach space valued functions*, Nonlinear Functional Analysis and Applications (Korea) (2017). accepted for publication
5. G.A. Anastassiou, *Principles of general fractional analysis for Banach space valued functions* (2017). submitted for publication
6. G.A. Anastassiou, I.K. Argyros, Equations on Banach space valued functions of abstract *g*-fractional calculus. J. Comput. Anal. Appl. **25**, 1547–1560 (2018)
7. I.K. Argyros, A unifying local-semilocal convergence analysis and applications for two-point Newton-like methods in Banach space. J. Math. Anal. Appl. **298**, 374–397 (2004)
8. I.K. Argyros, A. Magréñan, *Iterative Methods and Their Dynamics with Applications* (CRC Press, New York, 2017)
9. Bochner integral, Encyclopedia of Mathematics, http://www.encyclopediaofmath.org/index. php?title=Bochner_integral&oldid=38659
10. M. Edelstein, On fixed and periodic points under contractive mappings. J. London Math. Soc. **37**, 74–79 (1962)
11. J.A. Ezquerro, J.M. Gutierrez, M.A. Hernandez, N. Romero, M.J. Rubio, The Newton method: from Newton to Kantorovich (Spanish). Gac. R. Soc. Mat. Esp. **13**, 53–76 (2010)
12. L.V. Kantorovich, G.P. Akilov, *Functional Analysis in Normed Spaces* (Pergamon Press, New York, 1982)
13. G.E. Ladas, V. Lakshmikantham, *Differential Equations in Abstract Spaces* (Academic Press, New York, London, 1972)

14. A. Magréñan, A new tool to study real dynamics: the convergence plane. Appl. Math. Comput. **248**, 215–224 (2014)
15. J. Mikusinski, *The Bochner Integral* (Academic Press, New York, 1978)
16. F.A. Potra, V. Pták, *Nondiscrete Induction and Iterative Processes* (Pitman Publishing, London, 1984)
17. P.D. Proinov, New general convergence theory for iterative processes and its applications to Newton-Kantorovich type theorems. J. Complex. **26**, 3–42 (2010)
18. G.E. Shilov, *Elementary Functional Analysis* (Dover Publications Inc, New York, 1996)

Chapter 10
Numerical Optimization and Fractional Invexity

We present some proximal methods with invexity results involving fractional calculus. It follows [3, 4].

10.1 Introduction

We are concerned with the solution of the optimization problem defined by

$$\min_{s.t,\ x^* \in D} F(x^*) \tag{10.1.1}$$

where $F : D \subseteq \mathbb{R}^m \longrightarrow \mathbb{R}$ is a convex mapping and D is an open and convex set. We shall study the convergence of the proximal point method for solving problem (10.1.1) defined by

$$x_{n+1} = \underset{x^* \in D}{argmin} \left\{ F(x^*) + \frac{\gamma}{2} d^2(x_n, x^*) \right\} \tag{10.1.2}$$

where $x_0 \in D$ is an initial point, $\gamma > 0$ and d is the distance on D.

The rest of the chapter is organized as follows. In Sect. 10.2 we present the convergence of method (10.1.2) and in Sect. 10.3 we present the related application of the method using fractional derivatives.

10.2 Convergence of Method (10.1.2)

We need an auxiliary result about convex functions.

© Springer International Publishing AG 2018
G. A. Anastassiou and I. K. Argyros, *Functional Numerical Methods:
Applications to Abstract Fractional Calculus*, Studies in Systems,
Decision and Control 130, https://doi.org/10.1007/978-3-319-69526-6_10

Lemma 10.1 *Let $D_0 \subseteq D$ be an open convex set, $F : D \longrightarrow \mathbb{R}$ and $x^* \in D$. Suppose that $F + \frac{\gamma}{2}d^2(., x^*) : D \longrightarrow \mathbb{R}$ is convex on D_0. Then, mapping F is locally Lipschitz on D_0.*

Proof By hypothesis $F + \frac{\gamma}{2}d^2(., x^*)$ is convex, so there exist $L_1, r_1 > 0$ such that for each $u, v \in U(x^*, r_1)$

$$\left| F(u) + \frac{\gamma}{2}d^2(u, x^*) - (F(v) + \frac{\gamma}{2}d^2(v, x^*)) \right| \le L_1 d(u, v). \tag{10.2.1}$$

It is well known that the mapping $\frac{d^2(., x^*)}{2}$ is strongly convex [9]. That is there exist $L_2, r_2 > 0$ such that for each $u, v \in U(x^*, r_2)$

$$\left| \frac{1}{2}d^2(u, x^*) - \frac{1}{2}d^2(v, x^*) \right| \le L_2 d(u, v). \tag{10.2.2}$$

Let

$$r = \min\{r_1, r_2\} \quad and \quad L_0 = L_1 + \gamma L_2. \tag{10.2.3}$$

Then, using (10.2.1)–(10.2.3), we get in turn that

$$|F(u) - F(v)| \le \left| F(u) + \frac{\gamma}{2}d^2(u, x^*) - (F(v) + \frac{\gamma}{2}d^2(v, x^*)) \right|$$
$$+ \left| \frac{\gamma}{2}d^2(u, x^*) - \frac{\gamma}{2}d^2(v, x^*) \right|$$
$$\le L_1 d(u, v) + L_2 \gamma d(u, v) = L_0 d(u, v). \tag{10.2.4}$$

■

Next, we present the main convergence result for method (10.1.2).

Theorem 10.2 *Under the hypotheses of Lemma 10.1, further suppose:*

$$-\infty < \inf_{x^* \in D} F(x^*), \tag{10.2.5}$$

$$S_y = \{x^* \in D : F(x^*) \le F(y)\} \subseteq D, \quad \inf_{x^* \in D} F(x^*) < F(y), \tag{10.2.6}$$

the minimizer set of F is non-empty, i.e.

$$T = \left\{ x^* : F(x^*) = \inf_{x^* \in D} F(x^*) \right\} \ne \emptyset, \tag{10.2.7}$$

$$\| F(x^*) - x^* \| \le L_3, \tag{10.2.8}$$

$$L := L_1 + 2\gamma L_2 < 1. \tag{10.2.9}$$

Then, the sequence $\{x_n\}$ generated for $x_0 \in S^* := S_y \cap U(x^*, r^*)$ is well defined, remains in S^* and converges to a point $x^{**} \in T$, where

$$r^* := \frac{L_3}{1 - L}. \tag{10.2.10}$$

Proof Define the operator

$$G(x) := F(x) + \frac{\gamma}{2} \|x - x^*\|. \tag{10.2.11}$$

We shall show that operator G is a contraction on $U(x^*, r^*)$. Clearly sequence $\{x_n\}$ is well defined and since $x_0 \in S_y$ we get that $\{x_n\} \subseteq S_y$ for each $n = 0, 1, 2, \ldots$. In view of Lemma 10.1 and the definitions (10.2.8)–(10.2.11) we have in turn for $u, v \in U(x^*, r^*)$

$$|G(u) - G(v)| \le |F(u) - F(v)| + \gamma \left| \frac{1}{2} d^2(u, x^*) - \frac{1}{2} d^2(v, x^*) \right|$$

$$\le (L_0 + \gamma L_2) d(u, v) = L d(u, v) \tag{10.2.12}$$

and

$$|G(u) - x^*| \le |G(u) - G(x^*)| + |G(x^*) - x^*|$$
$$\le L d(u, x^*) + |F(x^*) - x^*|$$

$$\le L d(u, x^*) + L_3 \le r^*. \tag{10.2.13}$$

The result now follows from (10.2.9), (10.2.12), (10.2.13) and the contraction mapping principle [1, 5–8]. ∎

10.3 Multivariate Fractional Derivatives and Invexity

Let $X = \prod_{i=1}^{n} [a_i, b_i]$.

1. Let $0 < \alpha < 1$, we consider the left Caputo fractional partial derivatives of f of order α :

$$\frac{\partial^\alpha f(x)}{\partial x_i^\alpha} = \frac{1}{\Gamma(1 - \alpha)} \int_{a_i}^{x_i} (x_i - t_i)^{-\alpha} \frac{\partial f(x_1, x_2, \ldots, t_i, \ldots, x_n))}{\partial x_i} dt_i, \tag{10.3.1}$$

where $x = (x_1, \ldots, x_n) \in X, i = 1, 2, \ldots n$ and $\frac{\partial f((x_1, \ldots, x_n))}{\partial x_i} \in L_\infty(a_i, b_i)$, $i = 1, 2, \ldots n$. Here Γ stands for gamma function. Note that

$$\left| \frac{\partial^\alpha f(x)}{\partial x_i^\alpha} \right| \leq \frac{1}{\Gamma(1-\alpha)} \left(\int_{a_i}^{x_i} (x_i - t_i)^{-\alpha} dt_i \right)$$

$$\left\| \frac{\partial f(x_1, x_2, \ldots, x_{i-1}, ., x_{i+1}, \ldots, x_n))}{\partial x_i} dt_i \right\|_{\infty,(a_i,b_i)}$$

$$= \frac{(x_i - t_i)^{1-\alpha}}{\Gamma(2-\alpha)} \left\| \frac{\partial f(x_1, x_2, \ldots, x_{i-1}, ., x_{i+1}, \ldots, x_n))}{\partial x_i} dt_i \right\|_{\infty,(a_i,b_i)}$$

$$< \infty,$$

for all $i = 1, 2, \ldots, n$. Therefore, $\frac{\partial^\alpha f(x)}{\partial x_i^\alpha}$ exist for all $i = 1, 2, \ldots n$.
Now we consider the left fractional Gradient of F of order $\alpha, 0 < \alpha < 1$:

$$\nabla_\alpha^+ f(x^*) = \left(\frac{\partial f(x^*)}{\partial x_1^\alpha}, \ldots, \frac{\partial f(x^*)}{\partial x_n^\alpha} \right). \tag{10.3.2}$$

We replace in Definition 10.3.1 of Verma [9], $\nabla f(x^*)$ by $\nabla_\alpha^+ f(x^*)$.
2. Let $0 < \alpha < 1$, we consider the right Caputo fractional partial derivatives of f of order α :

$$\frac{\bar{\partial}^\alpha f(x)}{\partial x_i^\alpha} = \frac{-1}{\Gamma(1-\alpha)} \int_{x_i}^{b_i} (t_i - x_i)^{-\alpha} \frac{\partial f(x_1, x_2, \ldots, t_i, \ldots, x_n))}{\partial x_i} dt_i, \tag{10.3.3}$$

where $x = (x_1, \ldots, x_n) \in X, i = 1, 2, \ldots n$ and $\frac{\partial f((x_1, \ldots, x_n))}{\partial x_i} \in L_\infty(a_i, b_i)$, $i = 1, 2, \ldots n$. Note that

$$\left| \frac{\bar{\partial}^\alpha f(x)}{\partial x_i^\alpha} \right| \leq \frac{1}{\Gamma(1-\alpha)} \left(\int_{x_i}^{b_i} (x_i - t_i)^{-\alpha} dt_i \right)$$

$$\left\| \frac{\partial f(x_1, x_2, \ldots, x_{i-1}, ., x_{i+1}, \ldots, x_n))}{\partial x_i} dt_i \right\|_{\infty,(a_i,b_i)}$$

$$= \frac{(b_i - x_i)^{1-\alpha}}{\Gamma(2-\alpha)} \left\| \frac{\partial f(x_1, x_2, \ldots, x_{i-1}, ., x_{i+1}, \ldots, x_n))}{\partial x_i} dt_i \right\|_{\infty,(a_i,b_i)}$$

$$< \infty, \tag{10.3.4}$$

for all $i = 1, 2, \ldots, n$. Therefore, $\frac{\bar{\partial}^\alpha f(x)}{\partial x_i^\alpha}$ exist for all $i = 1, 2, \ldots n$.
Now we consider the right fractional Gradient of F of order $\alpha, 0 < \alpha < 1$:

$$\bar{\nabla}_\alpha^- f(x^*) = \left(\frac{\bar{\partial} f(x^*)}{\partial x_1^\alpha}, \dots, \frac{\bar{\partial} f(x^*)}{\partial x_n^\alpha} \right). \tag{10.3.5}$$

We replace in Definition 10.3.1 of Verma [9], $\nabla f(x^*)$ by $\bar{\nabla}_\alpha^- f(x^*)$.

3. Define for $k \in \mathbb{N} : \nabla_{k\alpha}^+ f = \nabla_\alpha^+ \dots \nabla_\alpha^+ f, k-$ times composition of left fractional gradient, i.e.,

$$\nabla_{k\alpha}^+ f = \left(\frac{\partial^{k\alpha} f(x^*)}{\partial x_1^\alpha}, \dots, \frac{\partial^{k\alpha} f(x^*)}{\partial x_n^\alpha} \right), \tag{10.3.6}$$

where $\frac{\partial^{k\alpha} f(x)}{\partial x_i^\alpha} = \frac{\partial^\alpha}{\partial x_i^\alpha} \dots \frac{\partial^\alpha}{\partial x_i^\alpha} f, k-$times composition of left partial fractional derivative, $i = 1, 2, \dots n$. We assume that $\frac{\partial^{k\alpha} f}{\partial x_i^\alpha}$ exist for all $i = 1, 2, \dots n$.

4. Define for $k \in \mathbb{N} : \bar{\nabla}_{k\alpha}^- f = \bar{\nabla}_\alpha^- \dots \bar{\nabla}_\alpha^- f, k-$ times composition of right fractional gradient, i.e.,

$$\bar{\nabla}_{k\alpha}^- f = \left(\frac{\bar{\partial}^{k\alpha} f(x^*)}{\partial x_1^\alpha}, \dots, \frac{\bar{\partial}^{k\alpha} f(x^*)}{\partial x_n^\alpha} \right), \tag{10.3.7}$$

where $\frac{\bar{\partial}^{k\alpha} f(x)}{\partial x_i^\alpha} = \frac{\bar{\partial}^\alpha}{\partial x_i^\alpha} \dots \frac{\bar{\partial}^\alpha}{\partial x_i^\alpha} f, k-$times composition of right partial fractional derivative, $i = 1, 2, \dots n$. We assume that $\frac{\bar{\partial}^{k\alpha} f}{\partial x_i^\alpha}$ exist for all $i = 1, 2, \dots n$.

5. Let $\alpha \geq 1$, we consider the left Caputo fractional partial derivatives of f of order α ($\lceil \alpha \rceil = m \in \mathbb{N}$, $\lceil . \rceil$ ceiling of the number [2]):

$$\frac{\partial^\alpha f(x)}{\partial x_i^\alpha} = \frac{1}{\Gamma(m-\alpha)} \int_{a_i}^{x_i} (x_i - t_i)^{m-\alpha-1} \frac{\partial^m f(x_1, x_2, \dots, t_i, \dots, x_n))}{\partial x_i^m} dt_i,$$
$$\tag{10.3.8}$$

$i = 1, 2, \dots n$. We set $\frac{\partial^m f(x)}{\partial x_i^m}$ equal to the ordinary partial derivative $\frac{\partial^m f(x)}{\partial x_i^m}$. We assume that

$$\frac{\partial^m f}{\partial x_i^m}(x_1, \dots, ., \dots, x_n) \in L_\infty(a_i, b_i)$$

i.e.,

$$\left\| \frac{\partial^m f}{\partial x_i^m}(x_1, \dots, ., \dots, x_n) \right\|_{\infty, (a_i, b_i)} < \infty$$

for all $i = 1, 2, \dots, n$. Note that

$$\left| \frac{\partial^\alpha f(x)}{\partial x_i^\alpha} \right| \leq \frac{(x_i - a_i)^{m-\alpha}}{\Gamma(m-\alpha+1)} \left\| \frac{\partial^m f}{\partial x_i^m}(x_1, \dots, ., \dots, x_n) \right\|_{\infty, (a_i, b_i)} < \infty,$$
$$\tag{10.3.9}$$

for all $i = 1, 2, \dots n$. Therefore, $\frac{\partial^\alpha f(x)}{\partial x_i^\alpha}$ exist for all $i = 1, 2, \dots n$. Now we consider the left fractional gradient of f of order $\alpha, \alpha \geq 1$:

$$\nabla_\alpha^{++} f(x^*) = \left(\frac{\partial^\alpha f(x^*)}{\partial x_1^\alpha}, \ldots, \frac{\partial^\alpha f(x^*)}{\partial x_n^\alpha} \right).$$ (10.3.10)

We can replace in Definition 10.3.1 of Verma [9] (or similar invexity formula), $\nabla f(x^*)$ by $\nabla_\alpha^{++} f(x^*)$.

6. Let $\alpha \geq 1$, we consider the right Caputo fractional partial derivatives of f of order α ($\lceil \alpha \rceil = m$):

$$\frac{\bar{\partial}^\alpha f(x)}{\partial x_i^\alpha} = \frac{(-1)^m}{\Gamma(m-\alpha)} \int_{x_i}^{b_i} (x_i - t_i)^{m-\alpha-1} \frac{\partial^m f(x_1, x_2, \ldots, t_i, \ldots, x_n))}{\partial x_i^m} dt_i,$$ (10.3.11)

$i = 1, 2, \ldots n$. We set $\frac{\bar{\partial}^m f}{\partial x_i^m} = (-1)^m \frac{\partial^m f}{\partial x_i^m}$ (where $\frac{\partial^m f}{\partial x_i^m}$ is the ordinary partial). We assume that

$$\frac{\partial^m f}{\partial x_i^m}(x_1, \ldots, ., \ldots, x_n) \in L_\infty(a_i, b_i)$$

for all $i = 1, 2, \ldots, n$. Note that

$$\left| \frac{\bar{\partial}^\alpha f(x)}{\partial x_i^\alpha} \right| \leq \frac{(b_i - x_i)^{m-\alpha}}{\Gamma(m-\alpha+1)} \left\| \frac{\partial^m f}{\partial x_i^m}(x_1, \ldots, ., \ldots, x_n) \right\|_{\infty, (a_i, b_i)} < \infty,$$

for all $i = 1, 2, \ldots n$. Therefore, $\frac{\bar{\partial}^\alpha f(x)}{\partial x_i^\alpha}$ exist for all $i = 1, 2, \ldots n$. Now we consider the right fractional gradient of f of order α, $\alpha \geq 1$:

$$\nabla_\alpha^{--} f(x^*) = \left(\frac{\bar{\partial}^\alpha f(x^*)}{\partial x_1^\alpha}, \ldots, \frac{\bar{\partial}^\alpha f(x^*)}{\partial x_n^\alpha} \right).$$ (10.3.12)

We can replace in Definition 10.3.1 of Verma [9] (or similar invexity formula), $\nabla f(x^*)$ by $\nabla_\alpha^{--} f(x^*)$.

Next we follow [4].

The discrete minmax fractional programming problem is

$$(P) \qquad \text{Minimize} \max_{1 \leq i \leq p} \frac{f_i(x)}{g_i(x)},$$ (10.3.13)

subject to $G_j(x) \leq 0$, $j \in \underline{q}$, $H_k(x) = 0$, $k \in \underline{r}$, $x \in X$,

where X is an open convex subset of $\mathbb{R}^n$ (n-dimensional Euclidean space), f_i, g_i, $i \in \underline{p} \equiv \{1, 2, \ldots, p\}$, G_j, $j \in \underline{q}$, and H_k, $k \in \underline{r}$, are real-valued functions defined on X, and for each $i \in \underline{p}$, $g_i(x) > 0$ for all x satisfying the constraints of (P).

Consider a function $f : X \to \mathbb{R}$ with fractional order derivatives, see [2]. In this section we give the definition based on the work [10] on several classes of generalizes convex functions.

Definition 10.3 ([4]) The function f is said to be $(\phi, \eta, \rho, \theta, m)$-invex at x^* for the left Caputo fractional partial derivative of order α, $\alpha \geq 1$ if there exist functions $\phi : \mathbb{R} \rightarrow \mathbb{R}$, $\eta : X \times X \rightarrow \mathbb{R}^n$, $\rho : X \times X \rightarrow \mathbb{R}$, and $\theta : X \times X \rightarrow \mathbb{R}^n$, and a positive integer m such that for each $x \in X$ $(x \neq x^*)$,

$$\phi \left(\left| f(x) - f(x^*) \right| \right) \geq \left\langle \nabla_\alpha^{++} f(x^*), \eta(x, x^*) \right\rangle + \rho(x, x^*) \left\| \theta(x, x^*) \right\|^m, \tag{10.3.14}$$

where

$$\nabla_\alpha^{++} f(x^*) = \left(\frac{\partial^\alpha f(x^*)}{\partial x_1^\alpha}, ..., \frac{\partial^\alpha f(x^*)}{\partial x_n^\alpha} \right). \tag{10.3.15}$$

Definition 10.4 ([4]) The function f is said to be $(\phi, \eta, \rho, \theta, m)$-invex at x^* for the left Caputo fractional partial derivative of order α, $0 < \alpha < 1$ if there exist functions $\phi : \mathbb{R} \rightarrow \mathbb{R}$, $\eta : X \times X \rightarrow \mathbb{R}^n$, $\rho : X \times X \rightarrow \mathbb{R}$, and $\theta : X \times X \rightarrow \mathbb{R}^n$, and a positive integer m such that for each $x \in X$ $(x \neq x^*)$,

$$\phi \left(\left| f(x) - f(x^*) \right| \right) \geq \left\langle \nabla_\alpha^+ f(x^*), \eta(x, x^*) \right\rangle + \rho(x, x^*) \left\| \theta(x, x^*) \right\|^m, \tag{10.3.16}$$

where

$$\nabla_\alpha^+ f(x^*) = \left(\frac{\partial^\alpha f(x^*)}{\partial x_1^\alpha}, ..., \frac{\partial^\alpha f(x^*)}{\partial x_n^\alpha} \right). \tag{10.3.17}$$

Similar concepts hold for $\nabla_\alpha^- f(x^*)$ and $\nabla_\alpha^{--} f(x^*)$.

References

1. S. Amat, S. Busquier, M. Negra, Adaptive approximation of nonlinear operators. Numer. Funct. Anal. **25**, 397–405 (2004)
2. G.A. Anastassiou, I.K. Argyros, *Intelligent Numerical Methods: Application to Fractional Calculus Studies in Computational Intelligence*, vol. 624 (Springer, Heidelberg, 2016)
3. G.A. Anastassiou, I.K. Argyros, S. George, Proximal methods with invexity and fractional calculus **27**(2), 84–89 (2017)
4. G.A. Anastassiou, I.K. Argyros, R.U. Verma, Role of fractional calculus in minmax fractional programming problems. Trans. Math. Program. Appl. **4**(2), 1–3 (2016)
5. I.K. Argyros, *Convergence and Application of Newton-type Iterations* (Springer, 2008)
6. I.K. Argyros, *Computational Theory of Iterative Methods* (Elsevier, 2007)
7. I.K. Argyros, A semilocal convergence for directional Newton methods. Math. Comput. AMS **80**, 327–343 (2011)
8. I.K. Argyros, S. Hilout, Weaker conditions for the convergence of Newton's method. J. Complex. **28**, 364–387 (2012)
9. R.U. Verma, Nonlinear demiregular approximation solvability of equations involving strongly accretive operators. Proc. Am. Math. Soc. **123**(1), 217–221 (1995)
10. G.J. Zalmai, Hanson-Antezak-type generalized $(\alpha, \beta, \gamma, \xi, \eta, \rho, \theta)$-$V$-invex functions in semi-infinite multiobjective fractional programming, Part I: sufficient efficiency conditions. Adv. Nonlinear Var. Inequal. **16**(1), 91–114 (2013)

Printed in the United States
By Bookmasters